Praise for *The Frugal Home~~

If anyone has put in his 10,000 hours and 10 years of experience, it's John Moody and family. That you and I get to glean from his experience and wisdom is a rare treat, a true honor.

—Joel Salatin, Polyface Farm, from the Foreword

This book is worth its weight in gold—actually, after a handful of years it's likely to be worth more. Unlike so many books this is an account of direct lived experience: jam-packed with tips and tricks from doing the homestead life each and every day. This would have saved me a lot of time if I had read it at the outset of my land endeavors. The resourcefulness shown in this work is impressive and though the warmer climate doesn't always apply to my locale directly, nearly all of the content is applicable to my situation with minor adjustments. Much of it is simply brilliant and makes me think of a variety of opportunities and approaches for adjusting my own system 15 years in.

—Ben Falk, homesteader and author, *The Resilient Farm and Homestead*

The Frugal Homesteader is a great resource for every homestead, proving that "affordable" doesn't mean "junk." Even as an experienced homesteader, John's storytelling pulled me in and kept me turning pages. He shares not only the great project results, but common pitfalls and problems to avoid.

—Laurie Neverman, creator, Common Sense Home

Just when I thought I had it all figured out, John Moody's book comes along. Wow! Suddenly it's time for a major re-fit. Not only is The Frugal Homesteader packed with well-thought-out innovations, it re-invents the word "frugal" to encompass the term "eco-friendly." This is a rich bounty of ideas that will be equally usefully to old hands and shiny new start-ups alike, whether on rural acreage or urban lot. It's written with an easy-to-read passion for the Earth and I totally love how this book rejects consumerism's need for new in favor of intelligent re-use and re-purposing. Delighted to have *The Frugal Homesteader* on my shelf! Thank you, John.

—Jenni Blackmore, author, *The Food Lover's Garden*
and *Permaculture for the Rest of Us*

I've been waiting for someone to produce a book that was worth my full endorsement and I've found it. Full of UN-common sense and overflowing with wit, wisdom and wonderful advice, *The Frugal Homesteader* is definitely going at the top of the list of "must own and read" books for all of my clients.

—Nick Ferguson, sustainable agriculture designer, consultant, and educator; and founder, Homegrown Liberty

In this hands-on book, John Moody gives readers the gift of practical advice they can use from the farm field to foraging in the forest. The result is a must-have for anyone looking to transition from homeowner to homesteader.

—Scott Mann, host, *The Permaculture Podcast*

John Moody turned "a run-down, rock-solid-clay, no-top-soil, 35 acres" into a productive, ecologically vibrant homestead. *The Frugal Homesteader* is chock-full of practical advice for doing the same, together with warnings about "gotchas" potentially disastrous for the unsuspecting beginner. "Frugal" to this author means more than saving money by, for example, utilizing salvageable resources otherwise going extravagantly to waste. He considers as well frugality of time and effort, with strategies that meet several needs in the same project—for example, managing pigs and chickens to forage natural feeds while making homestead-scale quantities of compost.

—Harvey Ussery, author, *The Small-Scale Poultry Flock* and themodernhomestead.us

THE FRUGAL HOMESTEADER

LIVING THE GOOD LIFE ON LESS

JOHN MOODY

new society
PUBLISHERS

Cover design by Diane McIntosh.
Cover images: © iStock Pig: 496988028 Coin: 697308644 + 533726165.
Text images: © Adobe Stock: p 1 goddessrising; p 4 Igor Serazetdinov;
pp 15, 89, 131, 177 blue67sign; p 51 ~ Bitter ~; p 111 valeriyabtsk.

Printed in Canada. First printing September 2018.

Inquiries regarding requests to reprint all or part of *The Frugal Homesteader*
should be addressed to New Society Publishers at the address below.
To order directly from the publishers, please call toll-free (North America)
1-800-567-6772, or order online at www.newsociety.com.

Any other inquiries can be directed by mail to

New Society Publishers
P.O. Box 189, Gabriola Island, BC V0R 1X0, Canada
(250) 247-9737

LIBRARY AND ARCHIVES CANADA CATALOGUING IN PUBLICATION

Moody, John, 1978– , author
The frugal homesteader : living the good life on less / John Moody.

Includes index.
Issued in print and electronic formats.

ISBN 978-0-86571-893-7 (softcover).—ISBN 978-1-55092-686-6 (PDF).—
ISBN 978-1-77142-282-6 (EPUB)

1. Self-reliant living. 2. Sustainable living. I. Title.

GF78.M67 2018   640   C2018-902683-9
                                                       C2018-902684-7

Funded by the   Financé par le
Government   gouvernement
of Canada   du Canada

New Society Publishers' mission is to publish books that
contribute in fundamental ways to building an ecologically sustainable
and just society, and to do so with the least possible impact
on the environment, in a manner that models this vision.

Contents

Foreword

by Joel Salatin

One of my favorite moments on farm tours is taking visitors to our cattle working chute in the barn and showing off the boards my dad and I scavenged from a tumble-down barn 50 years ago. Today we have a modern band saw mill and make lumber from logs we cut in the forest.

But back in the early days most of our lumber for building projects came from old barns dad and I tore down in exchange for the usable boards. We burned the junk stuff, but it had to be pretty junky to go in the burn pile. Today, we'd probably bring that home too and put in a hugelkultur bed. My mentor and *Stockman Grass Farmer* magazine founder, Allan Nation, always said that profitable farms have a threadbare look.

Here at our farm, Polyface, some 15,000 visitors a year drive down our lane and one of the most common remarks is "it's all so practical and plain." Our buildings are not flashy. Our infrastructure is functional. If we can slap some boards together and make something work, that's what matters; not whether we can adorn the front page of an agrarian lifestyle magazine.

When my wife Teresa was a junior in college majoring in home economics, she had to do a 10-year-out plan for budget, housing, and family. We were high school sweethearts (now we're old geezer sweethearts) and shared frugality from both of our families. For her project, she assumed we'd be living in an underground house, growing all our own food, making all our own clothes, cutting our own firewood—you know the drill. The monthly food budget for a family of four? A mere $200 (and most of that was toilet paper and facial tissue).

Teresa always made better grades than I did, so when this big project came back marked with a C– it was highly abnormal. The professor simply did not believe it was realistic. About 10 years later, my mom and dad moved out of the big farm house into a smaller house outside the yard. Teresa and I had our two children by that time and the attic apartment we'd lived in for 7 years was getting a bit cramped. As we cleaned out closets and began the move to downstairs, we found that old college project notebook and went through it.

Not only was it accurate, we were actually spending *less* than her supposedly unrealistic budget. We laughed and laughed. When we got married we had one car, a 1965 Dodge Coronet with 3-speed on the column I bought from a neighbor for $50. Drove it two years and sold it for parts for $75. How about that depreciation? In fact, when we'd been married 20 years we had not yet spent a total of $10,000 on automobiles. We never had a TV, never went out to eat, wore the same clothes until they fell off. We lived on literally a tenth of the income of all our college buddies.

But we were happy. We didn't buy toys for our kids. They were perfectly content to play with Tupperware on the kitchen floor. They made their fun and did not know their family lived at a third of the official U.S. poverty level. And we were content. In fact, most of our lives we've lived well below the poverty level. But who needs money when you have your own wood for heat, no air conditioning, don't have to drive to work, grow all your own food, and wear rags for clothes? And you're not in debt?

Since we didn't earn much we didn't pay income taxes. Take that, federal bureaucracy. We both came to this lifestyle because we both grew up in frugal families. Teresa remembers going out to eat once. I remember going out to eat three times. In the early 1960s my dad wanted a multi-use vehicle to avoid the typical farm truck and the car. He purchased a 1957 Plymouth sedan from a neighbor for $50 (apparently neighbors' $50 cars know they have a home here), stripped out the seat and doors, and used it as an all-purpose vehicle.

This was in the days before car inspection and mandatory seat belt laws. He inverted an old metal wash bucket for the seat (the original bucket seat) to sit on behind the steering wheel. Do you know how much room a 1957 Plymouth sedan has in it if you take everything out all the way back to the trunk? We hauled kids, chickens, pigs, calves, firewood, bags of feed—everything you can imagine in that old car. When car inspections came in he kept a set of tires for inspection and one to drive on. He'd put the nice tires on to get the car inspected, then come home and put the bald ones back on until they all went flat. That, my friends, is frugal.

To say I love John Moody's book would be the understatement of the year. I smiled all the way through it, marveling at the ingenious money-saving projects and reminiscing about my own growing up on a frugal farmstead. This book is chock-full of ingenious ways to do it cheaper, more efficiently, and perhaps most important of all, more child friendly.

I've always told farmers with big government cost-shared manure lagoons and other liquid waste management systems that I've never heard of a farm kid drowning or dying in a compost pile. And when the farmstead doesn't have a skull and crossbones door, you don't have to worry about the children getting into something toxic. I always say we don't put anything on the soil that you can't eat…at least in small quantities. I love the child-friendly thread woven through the fabric of this book.

Who needs video games and away-from-home entertainment when you have a cornucopia of plants, animals, meaningful projects and scientific marvels under your feet where you live? Creating a place of enchantment for youngsters might be the most ecologically sustainable thing we adults can do. That won't be a factory farm; it won't be a monospeciated orthodox chemical farm. It'll be a homestead menagerie of do-it-yourself trellises, tree houses, and teachable moments. Those don't take money; they take ingenuity and heart. Can you imagine anything more attractive to our children?

Some reading this book may think it's short on details. But I like that Moody stays with the big picture stuff. In fact, he uses the word stuff a lot, and that's exactly the way we talk to each other. We have far too many academic pontificates about things; what we need is a clear list of dos, don'ts, and how-tos from a dirt-under-the-fingernails long-time practitioner. If anyone has put in his 10,000 hours and 10 years of experience, it's John Moody and family. That you and I get to glean from his experience and wisdom is a rare treat, a true honor.

If you follow the advice in *The Frugal Homesteader*, you'll have a much greater chance of success. And you'll have children who take over chores instead of complaining about chores. Based on my experience, his advice is right on. Spend money on high quality tools. Buy bulk, always. Don't buy a tractor. Develop friendships. Grow it. Fix it. Build it. I guarantee you that our family's farming success today as a commercial outfit employing some 20 people full time is a direct result of generational frugality. We are now leveraging the benefits of being frugal.

I don't know if Teresa and I will ever really make much money, but we've often talked about what would happen if we did. Like if our ship actually came in. And we've decided we wouldn't live any differently. I'd still buy 50 cent shirts with somebody else's name on them; we still wouldn't go to Las Vegas; we'd still drive a cheap used car; we'd still have a basement full of canned goods from our garden. This is just a good way to live.

Thank you, John and Jessica Moody, for sharing your heart and how-tos with all of us who aspire to drop out of the high cash, fast cash, hedonistic rat race of hubris. I hope thousands of people resonate with this message and find in these pages the courage to embrace the ecstasy of frugal homesteading.

—Joel Salatin, Polyface Farms
PolyfaceFarms.com

Preface

Had you met me in my teens, you would have said, "He is never, ever going to be a homesteader or farmer." I had four food groups—sugary breakfast cereals, cookies, eggs (with sugar), and candy. I was a pasty-skinned, video-game-playing, cartoon-watching child of the '80s. I spent some time outdoors, generally only when my parents made me.

Had you met me in college, you would have said, "He is never, ever going to be a homesteader or farmer." I had eight food groups. I still played a lot of video games and watched a fair amount of TV, though I had become very active in sports as well.

Had you met me in my early twenties, you would still have said, "He is never, ever going to be a homesteader or farmer." That is, until I developed duodenal ulcers. Pain 24/7, like a small band of traveling dwarves, was mining my insides while holding a Metallica meth-fueled rave. Doctors could only offer me a lifetime of drugs. Instead, my then fiancée and now wife and I went with a radical change to our approach to food. We went from Kroger and Sam's Club to Wild Oats and Whole Paycheck. We graduated to the farmer's market, a CSA, and raw milk, and then to starting a food-buying club, the Whole Life Buying Club in Louisville, Kentucky.

At some point, in the midst of handling half a million dollars of local food each year and having kid after kid after kid, we thought, "Hey, wouldn't it be great to have a homestead in this mix?" I mean, with 2.5 kids, we weren't sleeping much anyway!

So, we started looking…and we looked, and we looked, and we looked, and eventually we ended up with 35 acres in the rolling hills an hour southwest of Louisville, Kentucky. We had no idea what we were looking for homestead-wise other than, "This is what we can afford, and this is how far from town we can be." We couldn't afford much. So we ended up with a run-down, rock-solid-clay, no-top-soil, 35 acres of semi-isolated beauty. In the first two years, we removed dumpster load after dumpster load of rubbish left by the previous owners and inhabitants.

We went to put in a garden. The ground broke the tiller before the tiller broke the ground. We adjusted to living 30 minutes or more from everything instead of three. Our closest town is just a few miles and has just a few thousand inhabitants. You don't reach a Walmart for 30 minutes in any direction. Even a trip to a local building supply store is generally an hour or more investment, requiring 20 minutes of driving each direction. The people in our area are good folks: mechanics, plumbers, and other tradespeople. But few farm, and even fewer homestead. Yet our neighbors appreciate what we are doing.

The homestead today.

Our place has changed some over the years since we settled. We rebuilt the barn and expanded it to provide a covered loafing area for larger animals and more storage. Some poorly built and dilapidated outbuildings went down, and a high tunnel went up. Pastures are slowly being renovated. Fencing will hopefully soon be replaced or upgraded. We continue to spread lovely soil across the rock-hard, exhausted clay landscape each season. Perennials, both domesticated and indigenous, now bear abundantly, providing food for man and beast.

I ended up a bit of both homesteader and farmer, depending on the year. As homesteaders, we try to raise as much as we can of what we need as a family—fuel in the form of wood, food through plants and animals, fertility through compost, water and whatever else we can self-produce—right here on the homestead. At the same time, we farm—that is, we seek to create an excess to sell to others. Some years it has been eggs and beef; others, pork and produce. As the farm and family change, we adjust and adapt as needed.

There is still so much to do around this place. Eventually, you begin to realize that homesteading is a journey you start but never finish. You build, you plant, you improve, knowing and hoping that one day, someone else will come along and see the value of what you have done, what you have contributed, and pick up the shovel or the hammer where you left them, and continue making the land a more beautiful and enjoyable, sustainable place.

Introduction

We should try to get results
with as little expenditure of time
and acreage as possible.
— Laura Ingalls Wilder,
"Economy in Egg Production," April 5th, 1915.

Ideas for the things at hand
to make our work easier will come to us
if we notice a little.
— Laura Ingalls Wilder,
"Shorter Hours for Farm Women," June 28th, 1913.

There is a movement in the United States today,
widespread and far-reaching in its consequences.
People are seeking after a freer, healthier, and happier life.
They are tired of the noise and dirt, bad air and crowds
of the cities and are turning longing eyes toward
the green slopes, wooded hills, pure running water
and health giving breezes of the country.
— Laura Ingalls Wilder,
"Favors the Small Farm," February 18, 1911.

Money in the Yard

A few years back, a friend gave me a copy of the Williams Sonoma catalogue. This particular issue was devoted to "Living the Good Life," with homesteading stuff scattered across the many pages. Boots. Shirts. Some tools. Chicken coops. Yes, a five-bird chicken coop for only $3,000 (delivery and assembly included). Now, if you have five chickens that lay 200 eggs each a year, and each egg is worth 50 cents, that is $500 a year in eggs. Of course, you also have costs like chicken feed, laying boxes and bedding, and all the other expenses that go into keeping chickens. But let's ignore that and all the labor, too. In six years you will finally break even on that coop.

Perhaps if you are a doctor or lawyer or some other high-dollar earner you can afford Williams Sonoma chicken coops. Or raised beds that run multiple hundreds of dollars a piece. Most of us dream of homesteading but wonder if we have the means to make it happen. How can we build a place that is beautiful without breaking the bank? How can we get by on less so we have time for more of the things we love and want to do with our lives? More time for nature, for nurture, for family, faith, and friends. Less financial stress. Less schedule duress. More fulfillment and freedom.

I hope this book will help you achieve your homestead dreams, and do so frugally and affordably. Not only will we share all sorts of ideas that have allowed our family—my wife Jessica, and our kids, especially Abigail, Caleb, and Noah—to homestead successfully even when we were without other sources of income, or during times of very low income, there will be lots of what we call "homestead hollers," ideas and innovations from friends and acquaintances across the land to give you inspiration and ideas to try out and adapt on your own place. Sometimes, no name is given to protect the people involved. You don't need to know someone's name to learn from their mistakes!

Please note that while this book is by no means exhaustive, all of the ideas are field tested. These are real solutions that we and others have used year after year, warts and all. I do hope you will find a lot of ideas that you can use, or that you can adapt to your situation, resources, and skills. Some of the projects will be more detailed than others—material lists and step-by-step guides so you can recreate a project at your place. Many more will be descriptive, giving you the basic idea and outline so that you can take what is and turn it into what could be, a thing of usefulness and beauty.

Most of all, I hope that reading this book will be inspirational—helping you see the endless possibilities for building an affordable homestead by making use of all sorts of low-cost or free resources that may otherwise go to waste or be overlooked by the modern world.

For much of our homesteading adventure, I have held other jobs at the same time as building our homestead, on top of going from two to five kids and enduring all sorts of other life changes and challenges along the way. Running a food-buying club in town, having multiple surgeries to repair an injury from childhood, taking a full-time job for over a year and so much more has interrupted or informed our family's efforts at homesteading. My hunch is many of you are just like our family—trying to balance a host of competing demands, things like work, family, hobbies, community, faith, and kids. The way we homestead seeks to keep all these areas in balance as much as possible.

So what is our approach to homesteading?

First off, "Never do anything that nature, an animal, a vegetable, or a microbe will do for you for free (and probably do better)." Homesteading is hard work at times. Why make it harder? Why make more work by not enlisting the natural allies just waiting in the wings?

Compost piles? Located close to growing areas and, if at all possible, uphill so that moving finished material is easier. Feed storage? Not far from the animals. Compost turning? Best left to the pigs, who will do it for a bag or

We employ chickens and pigs to turn our compost and control bugs as much as possible.

two of corn layered into the compost mix over time. Endless weeds and weeding? I would rather mulch once and then have just a few vigorous weeds to easily dispatch when they finally push through.

Either you and your family can work…and work…and work, or you can work with nature—animal, plant, and microbial—and enjoy all that free labor. Here is one simple way we do it.

A second principle that guides our approach to homesteading is "Work once; profit many times." Weeding is a never-ending battle, unless you decisively win the war with weeds. I would rather densely plant comfrey once along a fence edge to keep weeds out and also attract pollinators, improve my soil, and give me roots to sell, than mow those stretches every seven to ten days during growing season. Plan your ac-

 ## Abby's Lawn-to-Rabbit-Meat Tractors

I enjoy mowing our lawn. I enjoy it even more when our animals are the ones doing it. While chickens are problematic lawn mowers (their poop is very smelly if you step in it and they don't do a nice, even job of trimming the verge), rabbits make excellent lawn mowing crews. "Will work for clover," is my kind of motto! We have a pair of rabbit tractors that can be moved twice daily and that will keep a fairly large patch of front yard tidily trimmed. Instead of spending time and money keeping it mowed, each pass of the rabbits improves the quality of the soil with minimal time expense.

We hope to eventually move to four tractors and reduce my mowing to every few weeks and a few areas that are not appropriate for our four-legged herbivores.

Rabbits make good lawn mowers.

tions well and you will profit from them for months or years to come.

A third principle is that kids work too, so the place has to work for our kids. Infrastructure, tools, priorities, projects, and profits are all driven by our children—their interests, desires, skills, strength, and maturity.

Things are sized, weighted, and heighted for kid friendliness—buckets, hand tools, feed barrels. Even the dishes in our kitchen are located down low so that the littlest child can help set the table safely.

Throughout the book, you will see how this principle impacts a lot of projects and tools we have at our place. Speaking of tools, this is a good time to talk about what you will need to do the kind of stuff you will see in the coming chapters.

A Few Tools to Rule Them All: A Basic Tool Box

It doesn't matter how low cost a project is if you have to purchase expensive tools to make it happen. Almost everything outlined in the book can be done with the tools outlined below. Now, if you have certain tools or equipment not on the list, that can make projects easier or faster, but if you have the tools listed below (which, if you are homesteading or farming, you are going to need anyway), you should be good to go. If you are not homesteading yet, start collecting these tools now. If you hope to one day homestead, avoid the desire to acquire strange specialty tools—seeders, planters,

Small buckets and shorter, lighter hand tools make a big difference in the ability of small kids to help.

transplanters, tractor attachments, and what not—that you've never had practical experience with or may end up not ever needing or using.

Instead build a good basic tool set of high-quality implements that should last you many, many years. Properly cared for, some will last a lifetime and become gifts that you bequeath to your children. For my birthday my last year of high school, I remember my dad taking me to the store to get a high-quality set of starter tools. The mechanics tool set and the few additional items he picked out for me—some basic electrician's tools, a screwdriver set, vise grips, pliers, and a few other hand tools—have served me well for over 20 years. Many of the other quality tools that I have acquired have done over a decade's worth of work or more. Good tools are an investment well worth making and, if you are up to taking care of them, will be something you one day give to your children or their children.

Tool Rules

Buy the best quality you can afford for heavy-use tools—drills and saws, shovels and pitchforks, hatchets and mauls, and anything that will be used long and often around your homestead. Few things are as annoying as living in the middle of nowhere and having a tool break, delaying a project for hours or days.

Good hand tools are worth the cost and time to keep them up.

For single-use or rarely used tools, try to borrow or barter first. Then consider renting. Plan ahead on projects if you are going to need tools you don't have on hand. "Borrow, barter, rent" is my motto.

Try to avoid owning stuff that requires expensive specialty tools to maintain or upkeep. Some things are cheap up front because they cost a fortune to keep going.

For hoses, always get best quality. Low-quality hoses never last. If you find yourself needing water run to certain places all the time, it is time to put in more permanent storage or water lines or a higher-quality above-ground line if it is seasonal in nature.

Recommendations

I went back and forth for weeks trying to decide whether I should include recommendations for specific items. At the end of the day, my own experience, coupled with a friend's, convinced me to skip putting them in print. He recently purchased a handsaw from a company that previously had produced quality items across a wide range of products and had a sterling reputation. But unbeknownst to him, the company had recently changed hands, and the quality of their tools had dropped substantially and quite suddenly.

The tool he purchased lasted less than 20 minutes of moderate use. Even worse, the company had no intention of making it right with him. This happens more often than we like to admit, so instead of making recommendations in print that might no longer be reliable in six or twelve months, you can visit my website for more up-to-date information and recommendations. This way, I sleep better at night knowing no one is cursing my name as they deal with a dinky tool, and you can have greater confidence in recommendations I give.

Basic Tool List

Hand Tools

- Tape measure
- Basic mechanic tool set
- Vise grips
- Rope, especially smaller-diameter poly rope
- Quick square
- Metal snips
- Metal file set

Power Tools

- Mitre saw
- Angle grinder
- Cordless power tool set (drill, circular saw, and sawzall)
- Full set of bits for drill (wood, metal, adapters, and the like)
- Optional: skill saw, circular saw, table saw, chain saw

Mitre vs. Circular Saw

A friend suggested that homesteaders stick with a standard 7¼" cordless circular saw over the mitre saw. This isn't a bad suggestion, and if that is your go-to tool, I wouldn't quibble with you. But there are two reasons I stick with my mitre saw. First, my kids.

My kids use our mitre saw often and can do so very safely. Whether used for helping me with projects on the farm, cutting up certain types of firewood, or making forts and other creations, the mitre saw takes the brunt of our cutting calls. A circular saw takes a great deal more skill to use as safely as a mitre saw, though both require care and caution to use.

A mitre saw is great for cutting up certain types of firewood, such as slab wood and cutoffs, which we will talk about later in the book. Even better, it is a great way to teach basic tool safety to an age-appropriate child and get them involved in the work of putting up wood. I can't offer a set age, since children mature at such different rates physically, mentally, and emotionally. What some of our kids were ready to do at eight others needed until they reached ten or eleven. But with all our kids, the goal has been the same—to equip and enable them to contribute to our family and develop skills that will serve them well for the rest of their lives. Teaching kids to use power tools and then giving them projects and responsibilities to go with them not only makes our homestead life a lot easier, it makes it more fun for them and helps prepare them for the future.

Second, other tools I have on hand can take the place of a circular saw. In a pinch you can use a skill saw as a substitute for a circular saw, but not vice versa. *While it is best to have both*, I rarely use my circular saw, though often use my skill and mitre saw. A circular

With proper parental instruction, a child can safely use certain power tools.

saw is also generally easy to borrow. For longer cuts, a number of my neighbors also have table saws, which in some ways are circular saws on steroids. If you have money for only one, I would stick with a mitre saw and then a skill saw. Since most power tool kits come with a circular saw, most likely it will be a moot point, and you will have at least two of the three.

Garden Tools

- ▸ Garden cart
- ▸ Shovels
- ▸ Pitchforks
- ▸ Rake, heavy duty bow/garden
- ▸ Plant shears (save your kitchen or other scissors!)
- ▸ Optional: broadfork, leaf rake, hand spades

Give kids meaningful work and responsibilities around the homestead and the opportunity to develop skills while having fun at the same time.

HOMESTEAD HOLLER Amish Construction Crew

A local Amish construction crew reminded me of just how few tools you need if you know how to use them. A few years ago, I hired them to replace half the roof on my barn and also build a loafing area for larger animals on the backside. This involved installing six 6 x 6 uprights to support the new addition's roof. Instead of using a large circular saw or having to haul around a mitre saw for such cuts and run extension cords or generators, they brought and used a chainsaw for almost all lumber cutting. Occasionally, they broke out a handsaw for the two cuts that the chainsaw couldn't handle.

It showed me you don't need every tool under the sun, even to build a large building, if you know how to use the tools you have and use them well. Homesteader Matthew Eby says it well: "Many tools can be used outside their intended purpose. As a rule of thumb, I don't buy a tool unless I absolutely can't do the job without it, I know I'll be using it a number of times in the future, or it greatly reduces the frustration levels of getting a job or project done."

Wheelbarrow or Garden Cart?

Homesteading involves moving lots of heavy stuff. Compost. Mulch. Gravel. Wood chips. Firewood. Soil. Animal bedding mixed with manure. The list of what you will move is almost endless. A few years back, we calculated the amount of weight we moved by hand in a single year. It totaled over 300 tons. That is 600,000 pounds of stuff, or about 2,000 pounds per day! See why homesteaders don't need home gyms or gym memberships?

There are a number of ways to make all this work easier. But they are not all equal. First, don't bother getting a single-wheel wheelbarrow. They are not worth a nickel, save in limited work applications, almost none of which happen on a homestead. Don't waste your dollars on one. Whoever is using it also has to balance the weight it is carrying. Thus, they are not friendly to kids or adults, especially for heavier loads.

That leaves two options: double-wheel wheelbarrows and garden carts. For our first few years homesteading, we cycled through a wheelbarrow about every 18 months until it gave out. Then, a year or so back,

Even little ladies can move big loads with the right equipment.

we discovered the garden cart and haven't touched the wheelbarrow since. For our homestead, we purchased a standard garden cart, available at places like Tractor Supply or from Amazon. Some people choose the DIY route and build their own. Plans and designs are available online for free if that is your preference. With the garden cart we can move two to four times more weight than the wheelbarrow with half the work. It is more durable than the wheelbarrow as well and about the same price or less if snagged on sale. Even our smallest kids can help move the garden cart, while under moderate loads even our big kids struggled with the moving the wheelbarrow.

It is so easy to move that even when it is moderately loaded down, our fortyish-pound four-year-old can handle a hundred or more pounds of firewood without help and feel proud and like a helpful part of the family.

 **HOMESTEAD HOLLER   Saving on Tools**

Don't skimp on good tools, but do take advantage of sales. There are a few ways to save when tool shopping. First, ask about the floor models. Most stores have display models for many power tools. I happened to have a friend who was head of the tool department, who mentioned the store's clearance of these for a fraction of the normal price a few times a year. So I was able to purchase my first cordless tool set for over 70 percent off the normal price but still with its full warranty. Take a minute to ask the department manager at a few stores and see what deals you might find.

Second, tool sets often go on sale and clearance. I recently picked up a basic mechanics tool set for my son at a big box store for over 50 percent off. It was almost identical to the ones on the shelf 15 feet away, just clearanced so they could put a slightly different set in a slightly different box on the shelf. Looking back, I wish I had grabbed two or three of them, as my son's is now our house kit and saves me from having to run outside to pull tools from own main set.

Had I had a coupon, it would have been an even bigger savings. Combining a sale with a coupon at a place like Tractor Supply or Home Depot can save you 30 to 40 percent off a tool's best price. Make a list of what you need, get a good idea of what the going price is, and then keep an eye out for when conditions are right to add it to your collection.

HOMESTEAD HOLLER — Saving on Shop Supplies

Sanding pads. Cut-off wheels. Sawzall blades. Nails and screws. A lot of the expense of DIY projects isn't the tools themselves but the bits and blades and such other stuff that tools need to get their jobs done. If you go to the average big box store, some of these items are fairly expensive. A cut-off wheel can set you back four bucks, and a steel barrel may take two wheels to cut in half for a simple project. That adds almost ten dollars to a project before it is really under-way! What if you could get 30 cut-off wheels for the price of three? This is where online retailers excel. There are a host of bulk items that every farmer and homesteader needs that online retailers supply at a fraction of the cost of big box stores. Even as semi-remote as we are, these items are delivered right to our door. Also, these items, as long as stored properly, never expire or otherwise go bad. So there is no risk in stocking up to save.

Also, avoid the temptation to multiply screw sizes. Instead choose two good sizes that let you do the bulk of the work around your place, and buy these two sizes of screws in *bulk*. Generally, the larger-size containers of nails and screws cut the cost per screw in half over the small boxes that people grab off the store shelves. I stick with the Torx-style heads and suggest you do the same—they are slightly more expensive and worth every penny in terms of stopping stripped-screw aggravation and lost time from dealing with stripped and broken screws.

A bucket of 2½-inch and a bucket of 3½-inch screws has been almost all I ever needed over 95 percent of the time when doing projects. On oc-casion, when I absolutely have to have a different or special size, I will get a small container unless I know I will need more. Otherwise, we make the above sizes work.

Given that we live in the middle of no-where—even a trip to our local building supply company eats about an hour of time—the time and money savings are substantial when you stock up. Also, I have found no difference in quality over many years of ordering items in bulk online versus what big box stores had to offer. In terms of making farming and homesteading affordable and bringing down project costs, the 50–80 percent we save in these areas lets us invest in other things that enrich not just our farms and homesteads but our communities.

Given how many of these we go through in a year, bulk buying results in big savings both in cost and time.

Holidays, Birthdays, and Achieving a Homestead

If you have kids, or friends with kids, let me make a suggestion. Most kids have far too many toys and almost no tools. By the time a child reaches their teens, or better, even earlier, stop buying them toys and start giving them tools that will keep their value and be of use the rest of their lives. A great gift to a teen is a basic mechanics tool set. Each year after that you can add on additional tools. A seven-year-old may be ready for a good pocket knife. An eight-year-old should have a good, basic bow. A nine-year-old may benefit from a multi-tool. By twelve, basic tools should be on the birthday table. By the time they reach their late teens, our kids should have all sorts of basic items needed to manage and maintain a house and homestead, both in terms of skills and stuff. Long after Legos and other toys have been discarded, sold used at a discount, or dustbinned, these gifts will continue to bless and build up your kids and your family.

[1]

In the Garden

If you have a garden and a library,
you have everything you need.
— Marcus Tullius Cicero

⸙

Odd as I am sure it will appear to some,
I can think of no better form of personal
involvement in the cure of the environment
than that of gardening. A person who is growing
a garden, if he is growing it organically,
is improving a piece of the world.
He is producing something to eat,
which makes him somewhat independent
of the grocery business, but he is also
enlarging, for himself, the meaning of food
and the pleasure of eating.
— Wendell Berry

⸙

A society grows great
when old men plant trees whose shade
they know they shall never sit in.
— Greek proverb

⸙

One of the big reasons people want to homestead is so that they can grow their own food. They want space for a garden, a real garden, not just some corner of a shady back yard or a small plot at the community center. They (wisely!) want to cut down their food bill, which for a homesteader is one key to long-term success. Unfortunately, gardening can easily become an expensive adventure, especially for first timers.

This section is all about ideas to help your garden gloriously grow gobs of food without requiring great amounts of green—$$$—inputs.

Small Starts Lead to Success: Garlic in Boxes

How do you grow when you have nothing to grow in? When we moved out to our farm we had a problem. We had no dirt. None at all. We were greeted by rock-hard clay, so tough that even the tiller I borrowed from a neighbor couldn't take it. I was determined to grow something no

 **Easy Does It**

For some reason, when people are in trouble, they tend to call you on a Friday night. Late on a Friday night. At least they seem to do this to me. One balmy late July evening, a bit after nine o'clock or so, my phone rang. I answered and was surprised to hear the voice of Sam, a friend whom I hadn't seen much of lately. (We live about 80 miles from each other.) "John, I need to talk to you about a problem...a gardening problem." Before you laugh, his voice was quite serious. His situation helped make sense of why.

He and three families from his church had decided to create a community garden. Chickens. Bees. Beautiful vegetables. Fresh food all summer and fall long with extras for canning and preserving and putting up for winter—at least, that is what they thought. One family had land, and together the group set about putting in their first garden. A 10,000-square-foot garden. Now, if you have a fair bit of experience and the right equipment, going straight to 10,000 square feet (about a quarter of an acre) isn't a terrible idea. But if you are new, and if you don't have a good game plan, especially for dealing with weeds and nature's other misdeeds, well... you end up like my friend, making phone calls late at night to other growers hoping for some miraculous deliverance.

Their 10,000-square-foot garden had started off pretty typically. Amend, till, plant. Bee hives were bought and built. A chicken coop was framed and filled with various cluckers of all

sorts of shapes, sizes, and colors. They invested a great deal of time and, from the sound of it, a decent sum of money as well. In a few weeks, things looked lovely—until the weeds came. Soon after, so did the deer. Not to mention the raccoons, possums, and everything else that enjoys a chicken dinner with a few servings of fruits and vegetables on the side. Animals care about a balanced diet, too.

What happened to the plants that survived? Answering that involved looking, and looking, and looking, because you couldn't find it. Two-foot, three-foot, four-foot weeds filled the entire garden. As things started to get away from them, they tried to till again. They weeded. The weeds bounced right back. An emergency meeting was called. All four families would spend an entire weekend saving the garden! Sam didn't think this was a good plan. Five hundred hours of labor to save maybe a thousand or so dollars of vegetables, at best. The math just didn't make sense.

If you learn anything from my friend's story, it's to not overdo it out of the gate. Start small. Take on a few projects at a time. Create a few-hundred-square-foot garden. Master a couple of crops each season. In a few years you will be so much further along than if you had burned yourself out going too big too soon and too fast. Homesteading is like a marathon, no one wakes up one morning and runs 26 miles. The first guy who tried it, back in ancient Greece, finished the run but died a few minutes later. Take it as a cautionary tale. Build up your stamina, skills, and homesteading sense. One day you will get to where you want to be and you will enjoy the journey so much more along the way.

Small beginnings done well set the stage for big success later.

matter what that first fall, and the only quick solution I came up with was growing garlic in cardboard boxes. If I couldn't grow in the soil, I would do it on top of the soil until I had soil of my own to work with!

The experiment was a moderate success and also resulted in a nice small area for our first full year of gardening the next spring. I didn't realize it at the time, but it was a type of container growing meets lasagne gardening, just using boxes instead of longer-lasting plastic-type totes. Most of all, it allowed us to have some success right from the start. Don't underestimate how important doing a few small things well is the first few years on your homestead. Too many early setbacks may send you straight back to the city.

A few notes about growing in boxes. First, if you grow in boxes, make sure you lay a few broken-down boxes under and around the growing box so that, come spring, grass and weeds aren't growing right

John's Uncommon Two Cents—Safely Using Cardboard

Remember, with cardboard boxes, to remove any tape, staples, or other metal clasps, and any shiny stickers or similar stuff on the boxes. I don't bother with boxes that have the nylon-style tape (strings inside the tape). That stuff is too much of a pain to remove to be worth the time. Stick with low- to no-ink boxes as much as you can. While the inks are supposed to be mineral- or plant-based and safe, I still feel better exposing my soil to as little of them as possible, including the newer soy-based ones.

Look for cardboard that has little to no ink, along with easy to remove tape and staples if any.

up against your raised boxes. Second, if you are in a really cold climate, mulching around the boxes with wood chips or similar material is recommended. Raised boxes standing alone outdoors can get so cold that the garlic suffers damage. Mulch—both around and on top of the boxes—helps protect it from cold, similar to when the garlic is planted directly in the ground. By the time the garlic is ready to harvest, the boxes will be completely broken down and the soil under and around significantly improved. Third, plant cold-hardy varieties, especially the farther north you find yourself.

Perennial Cardboard Plant Protectors

Perennials often get planted in places surrounded by grass that gets mowed. Unfortunately, this means some of our perennials have gotten mowed down over the years by neighbors lending a hand and mowing our lawn, or even by me not noticing them amidst tall grass I got behind on keeping in check.

So, I came up with a simple solution to protect new transplants—I plant per normal instructions but place a cardboard box around them.

This makes it very clear to anyone, including children playing around the property, where the plants are while still small, providing a fair bit of protection until they become larger and well established. The boxes also hold freshly applied amendments, soil, and mulch in place while things get settled and established. They are easy to weed and also easy to inoculate or populate with a desired companion plant such as clover.

Expensive perennials made a bit more safe from mowers and small children.

Easy Raised Beds with Slab, Scrap Wood, Straw Bales, or Concrete Blocks

Raised beds are a popular way to start growing in small spaces. They can also be very expensive. Materials for an average-sized raised bed can run a few hundred dollars. Like the Williams Sonoma chicken coop,

On a gentle slope, slab wood on the lower side makes for easy raised beds and a beautiful setting to grow food.

that makes for a much longer return on investment. But you can make beautiful and affordable raised beds out of a lot of materials, both low-cost new or recycled.

Slab-wood or "second"-wood raised beds are especially useful if your land has a gentle slope. You can make a path, swale, and raised bed system that is fruitful, water retentive, erosion resistant, and easy to maintain.

Using brand-new lumber is expensive. Instead, we visit our local sawmill and ask if they have any forgotten or rejected lumber for sale. This lumber is often 50–80 percent less, yet for our purposes suitable for many years of service. We look for boards that are at least six and ideally eight to ten inches in width, a minimum one and a half inches in thickness, and eight feet or longer in length. If there are nice and well-priced wider boards, we wouldn't walk away from those either.

To hold the raised bed boards in place, you have a few options. Some people use rebar. If you have access to free or low-cost rebar, this isn't a bad idea. Since I have access to lots of low-cost wood, I make wood pins to hold the boards in place. All it takes is the mitre saw set to a 30 or so degree angle and the wood. Decide on the length of pin you need, and cut away. Leftover pieces make good plant row markers or kindling for fires.

All it takes is some scrap 1×1 wood and a mitre saw to make simple wood pins to hold short raised bed walls in place.

Either way, your pins should be double the height of the board. So if the board is 8 inches, your pins should be 16 if the depth of your soil allows. The looser your soil, the longer you should make your pins. Pins generally last me about two to three growing seasons and then get replaced.

Rebar pins are easier and more durable than wood ones, but make sure you drive them a inch or so below the top edge of the board. A rebar pin sticking up out of the ground is a real danger if someone falls on it while working in the garden. The wood pins are only slightly less so, so drive them at or below the board's height as well. You can also use tree branches, especially locust or cedar because they should last longer than many other species, though any hardwood would do.

Straw Bales and Concrete Blocks

Wood isn't the only way to make a raised bed. You can also make a raised bed by using straw bales or concrete blocks. Bales provide great insulation and moisture retention, while also giving plant roots another place to spread into (some people grow directly in straw bales, but that requires a lot of bought fertilizer, so it is not something I recommend). The added benefit of using straw bales to make raised beds is that you have mulch right next to where you need it. As the season moves along and the bales begin to break down, you can bust them open and spread them to suppress weeds since the soil should now be warm and stable from settling and all the plant roots taking up residence in the growing space. Just see our note below about safely sourcing straw.

Straw bales also make quick, easy, and inexpensive raised beds.

If you are reusing concrete blocks, clean and rinse them thoroughly, and make sure there are no chemical, paint, or other unknown and possibly dangerous residues on them. Remove any mortar or other leftover debris that is on the blocks as well.

Some people get creative and even fill the upright channels of the blocks with soil and do flowers or similar plants for a lovely border on their concrete raised beds. Another option is to fill them with rock or whatever other substance you won't mind having in that spot if you ever take the raised bed apart. Also, consider whether if you will ever expand the bed as well when you fill the channels, as moving the blocks is going to make some amount of mess as the content of the channels and the bed itself spill out.

The Mess of Hay, Straw, Manure and Compost

One of the best ways to get soil for your first garden or to recharge your current one is through manures or compost. Unfortunately, these low-cost and high-value fertilizers are now also one of the greatest dangers to your garden. As herbicide resistance has swept across the United States and other places, new herbicide formulations have flooded the market. While the old versions were a temporary, highly damaging danger to your growing spaces, these new ones are a multi-year disaster if your growing spaces get exposed to them.

One of my friends found this out the hard way. He had a decades-old organic garden in Florida. To amend it, he purchased a few pickup truck loads of manure. Not much at all given the size of his garden. Early spring, the loads arrived. He spread and worked them in as usual. Then he planted and transplanted his crops. A few weeks later it was clear something was very, very wrong. It took some time, but eventually he found out that he had purchased contaminated compost.

Herbicide contamination is something you need to take very seriously. Unfortunately, pyralids (a chemical component of some modern herbicides) and many other herbicides may be found in manure, compost, straw, and hay. If you are bringing any of these onto your farm, you need to source them carefully to protect your crops.

With the pyralids, if you get them in your garden, it is usually three to five years before you can grow much of anything other than corn. If you are Certified Naturally Grown or Organic or some other certification, it will be lost as well for quite some time. Again, if you are bringing fertility or hay and straw onto your place, make sure you source them carefully. A small mistake can set you back many, many years, and many thousands of dollars. Just ask my friend or the tens of thousands of other growers who have been hit by herbicide contamination.

Fencing and Mulching with Sheet Metal

Most people don't think of sheet metal as a garden input, but it has a lot of interesting uses. First, it makes a great mulch. Yes, mulch! While grass and weeds will still push through wood chips and straw, nothing can pass metal mulch. So if you have spots that have really bad weeds you need to suppress and you don't want to mess with constant hand or other controls, metal mulch is a great option. Also, unlike other mulches, it will last as long or longer than you. Plastic rips, tears, degrades, and needs to be removed. Cardboard and newspaper decompose. Wood chips and straw need replacing. But once you lay sheet metal down, it will stay for ten to twenty years or more.

Also, its weight to coverage ratio is great. Most sheets of roofing and siding metal are three feet across and ten or so feet long, so an average sheet will mulch 30 square feet. With help from one of my kids, we can cover 300 square feet in 15 minutes or less. With wood chips or some other mulch, we would need thirty or more minutes to cover the same area, and would still have some weeding and other work to do later on. I personally like to use metal mulch for edges and other areas that are hard to keep weed free, along with paths if I run out of sufficient organic mulches or if a particular path or area has some sort of persistent, troublesome pest to deal with weed wise. As you will see below, some growers go full-metal garden, making every path metal.

Not only is sheet metal a great mulch, it also makes excellent fencing. Unlike wire, which rabbits and other flexible critters can easily

HOMESTEAD HOLLER The Sheet Metal Garden

Herrick Kimball's website was what first inspired me to make use of metal mulch many years ago. His neighbor, Steve, was featured in a few of Herrick's posts on how he created a low-maintenance, high-yielding garden. Metal mulch played a crucial role both in weed suppression and pest exclusion. I reached out to him to see if he would take some updated pictures to share here. I also asked him about pests. One problem in my area is that the metal attracts some predators, like snakes, but also lots of pests, especially mice (though so does straw mulch). Herrick blogs at thedeliberateagrarian.blogspot.com.

Some people are now using sheet metal as yet another way to build raised beds. At around three feet high, these are especially convenient for older people or those with back problems. The only drawback is that three feet deep is a lot of space to fill with soil. I have seen people use straw, hay, or hugelkultur—the practice of burying wood to build soil—in the bottom to cut down on the cost and time it takes to fill the raised beds and to help build high-quality soil over time.

Credit: Herrick Kimball

Sheet metal is a permanent approach to mulching, stopping weeds and grasses for as long as you leave it in place.

Scrap sheet metal makes a decent fence to exclude a number of common garden pests, such as raccoons, possums, and rabbits.

pass through, sheet metal keeps almost everyone out of your garden. Also, for larger pests, wire is pretty easy to climb (my melons have been multi-year victims even inside our fully fenced garden). Sheet metal is slippery and offers no easy places for climbing critters to claw on to, so it is also effective against raccoons, possums, and other agile garden attackers. To help prevent burrowing, you can make a thin trench and place the wire two to three inches into the ground and then replace the soil, or lay chicken or similar wire underneath, extending twelve or so inches out on all sides.

Making Sheet Metal Safer

Sheet metal can be very dangerous, especially along its edges and corners. A simple way to make sheets safer on chicken and rabbit tractors and other creations is by bending the corners over with a mallet or hammer.

This protects people big and small from slicing themselves badly along the corners. Sheet metal used as mulch should be secured in place by at least two concrete blocks or large rocks. Unsecured sheet metal is the equivalent of a flying guillotine. It can damage both buildings and bipeds, so let's make sure that doesn't happen on our homesteads. Sheet metal fencing should be thoroughly secured to T-posts by drilling holes in the sheet metal every 12 or so inches and then using wire or some other method to secure the metal fence to the posts. Don't use twine or rope, either natural or synthetic. The natural swaying of a fence because of wind and weather will cause it to fray and then break quite rapidly.

Mineral Tote Growing Bins

One of our favorite 2017 experiments was using mineral totes—large plastic containers that hold mineral supplements fed to animals—to start transplants and grow stuff throughout the season. In rural areas, farms with large animals use a fair number of these each month, especially if the farm operation is large. Instead of a one-way trip to the landfill, we collect them and use them for all sorts of farm projects—storage, worm composting, soil block mixing, and container growing, among many other options.

All you need to do to make use of these totes is ensure they drain properly. A drill with a small-diameter bit (¼-inch or slightly larger) will allow you to put 20 or so holes in the bottom, providing adequate drainage and air exchange. It is good to layer the bottom two or so inches of the totes with rotted wood chips or similar compaction-resistant material to ensure that the holes don't become clogged.

These bins produced around 30–40 pounds each of sweet potatoes. One bin produced so many sweet potatoes it split clean in half! You can also use them for a more traditional container-type garden, where you use two stacked on top of each other, with the lower serving as a water reservoir.

Many larger-animal farms use lots of food-grade mineral totes that can serve many purposes on a homestead.

You can grow almost anything anywhere if you really, really want to.

If you go this route, I wouldn't do such a heavy, productive root plant as sweet potatoes in them but instead focus on above-ground or other lighter crops.

Growing Up Instead of Out:
Plant Cages, Trellising, and More

"Grow up!" No, I am not insulting you. Many crops that tend to sprawl can be grown vertically to conserve precious garden space, make harvest easier, and protect them from pests and disease. Cucumbers, some melons and squashes, and many other plants can go up instead of out, increasing your yield per square foot substantially. Even some surprising, space-hogging plants like sweet potatoes can be grown up to conserve space.

When you have limited horizontal growing space, growing up can allow you to double your yields per square foot, if not more. You will need to be more mindful of water and soil fertility, as a densely planted garden consumes a great deal more of both. On the plus side, the more densely planted, the less space for weeds and other misdeeds to take up space. Smaller spaces also take less time to water and take less work to manage.

Trellising is what this approach is usually called. There are many ways to do it, so let's look at some of my favorites and the materials you will need.

Concrete Reinforcing Wire (CRW)

If you are looking for a simple way to make strong, durable cages for supporting plants, concrete reinforcing wire works exceptionally well. I was introduced to this idea by an urban homesteading friend named Paul Brown, who kindly gave us blackberry cuttings and a number of other plant starts many years ago when we were first starting our homestead. When I went to dig the volunteer plants that he was just going to mow away in his urban food forest, I noticed a garden full of cages containing all sorts of plants—peppers, tomatoes, blackberries, and more. These simple cages were made from rolls of concrete reinforcing wire.

Simple and durable plant trellising.

The rolls are significantly less costly per linear foot than the flat sheets of CRW and, while heavier, are easier to transport.

There are a lot of ways to cut CRW. I prefer using bolt cutters or an angle grinder. Don't use TIN SNIPS/metal cutters. You will ruin the blades. (Don't ask me how I know this.) Treat your aviation and metal cutting shears like your wife treats her sewing scissors— save them only for their appointed task!

Material List

- ▸ Roll of concrete reinforcing wire
- ▸ Pig pliers and nose rings

If you are doing this job solo, make sure you grab four or so concrete blocks to keep the concrete wire laying flat while you cut it and to keep it, when cut, from flying up and hitting you. It will naturally want to roll back up, which makes finishing the cages very easy, but may make cutting the first few—when you have the entire tension and weight of the roll working against you—tricky until you get the hang of it.

Roll out a section of the CRW, setting two blocks on the edge and then two next to the roll. You should have around 12–14 squares fully rolled out. We use cages that have ten full squares (60-inch circumference), so we cut every eleventh row. After the cage section is cut free, allow it to form back into a circle. Using the pig pliers and rings, at the top, middle, and one section above the bottom, secure the two sides together. At this point, you have a few options with the bottom. You can cut out "feet" by lopping off the outer edge of the cage, leaving roughly six-inch-long spikes that make the cages far more secure for heavy crops or in high winds. Or you can leave them as is and secure them some other way.

These cages have been our go to for years for a wide variety of fruiting plants, especially tomatoes, peppers, cucumbers, peas, and sometimes pole beans, among others By growing these up, you improve yields, reduce disease and certain pest pressures, and make harvest much easier. You can also use them with non-traditional crops like sweet potatoes if you are really squeezed for space and need to keep them from sprawling and overtaking your entire growing space. So if you drop by our place in summer you may see some strange things sprawling upwards towards the sun instead of spreading across the ground.

By cutting off the final cross wire, you can make the cages self-staking and use them to protect young plants that sometimes go unnoticed in the garden.

The CRW provides excellent support for plants while not impeding the ease of harvesting large produce such as full sized peppers, tomatoes, cucumbers, and more.

The cages last anywhere from five to ten years, depending on your weather and approach to storage, bringing your annual cost per cage down below a dollar for plants that will hopefully produce 20 to 50 times that value each growing season.

Safety First

It is generally best to work in a team on the homestead. But sometimes you find yourself needing to get a job done solo. The above trick of using some extra blocks to hold the wire so I could safely cut it came about from one of those occasions. Sometimes, safety takes as little as a few seconds of moving some extra stuff lying around. Always take the time for it. It is a small cost in time for a big savings in possible present and future pain!

Turn Your Homestead Work into Wages

A few years ago, I realized that, with so many people getting into gardening and homesteading, you can often sell stuff like plant cages at a tidy profit. So I would take extra orders for cages and make a bunch at a time each spring for sale. Upper-scale folks might prefer the look of welded wire to concrete wire, so this is something to consider. We do the same with worm compost—since we are going to make it anyway, why not make some for sale since the small additional time results in a good business opportunity for my son? Such things can be profitable little side gigs or great ways for kids to make spending money if you have the market.

Pallet Propagation and Drying Tables

Plant starts are usually placed on tables to make care, watering, and other work easier. Pallets are a great way to make these tables. The same design used in spring for starting plants can then in summer and fall be repurposed for saving your harvest. For us, this requires simply laying a cattle panel or wire across the table to create a space and additional support to lay the drying produce on. If you have rabbits and design the table right, you can then get a third use out of it by relocating some of your rabbit hutches onto the table come late fall and over winter. Since our table is in our high tunnel, this gives the rabbits a slightly more comfortable environment for winter while also dropping fertility where it will soon be needed instead of having to move it.

 **HOMESTEAD HOLLER** Beautiful Cattle-Panel Arched Trellis

Another neat way to grow up involves using standard cattle panels. A fair number of homesteaders now use such arches in various types of fruit and vegetable growing. In the spring these structures allow full light onto the early plantings—lettuces, cabbages, and other early-season, cool-loving crops. As spring turns to summer, heat-loving crops like beans, cucumbers, melons, and more are planted along the edges and trained up the trellis. As summer hits full steam and the vining plants are trained across the trellis, it provides much needed shade relief for the plants below that don't like the full sun and high temperatures. Inside the trellis, temperatures can be 10–20 degrees lower than out in the open. This allows lettuces and many other plants to enjoy many weeks longer of harvest.

Many crops can be grown on the arched trellis—beans, lighter varieties of winter squash, cucumbers, smaller melons, and more. The best part is that after summer use the structure can be covered with a tarp or plastic and used as animal housing or turned into a greenhouse for late-fall and winter production. The structure also isn't too difficult or heavy to move—a four-wheeler, riding lawn mower, or a pair of strong people can relocate it.

Both concrete reinforcing wire and even welded wire can be used to make smaller arches for trellising similar to those made from cattle panels. While less expensive, they cannot hold anything near the weight of a cattle panel, so they are best used only for green beans, cucumbers, peas, and other light crops that won't cause a possible collapse.

You can use many different materials to make propagation and other types of work tables. We used some scrap lumber, a few pallets, and a cattle panel.

Material List

- ► 2 or 3 pallets (depending on size of table you plan to create)
- ► Scrap lumber, 3 pieces to table length
- ► Screws
- ► Cattle or hog panel, or welded wire or similar material

The ideal height for these depends on a lot of factors. Do you want them to be multi-use? Single or double layers? The easiest way to build these is, first, start with a level spot. Cut your pallets down to the desired height (1). Decide on the length of the table and cut your scrap lumber to length to make the rails (2). We go with about nine feet to match our cattle panels. Lay out the pallets (3A and 3B) and secure the rails across the pallets. Then, lay the cattle panel across the top (4). Your table is now ready for plant propagation duties.

Compost Underneath The Tables

Plant starts are sensitive folks. Tiny plants dislike changes in temperature, especially sudden cold temperatures that are common during late winter and early spring, when you are trying to get a head start on your garden by starting it inside. But providing heat, especially at night, can be an expensive enterprise. If you are starting plants outdoors in a high tunnel or similar structure, why not provide heat for free through the power of compost?

This approach has a long and rich history. Back in France during the 1700s and 1800s, market gardeners would scurry around the city collecting horse poop to replenish their soil and provide warmth to their plants. Their system was so successful, they supplied this large city with much of its produce *year round*. The heat did many things for the plants, like improving germination, protecting from cold and frost, and allowing the farmers to create customized micro-climates best suited for each crop.

We can no longer recommend composting in the same space as your plant starts, because of the overuse of antibiotics, especially in animal agriculture, and other food-safety issues. If you are in any way selling produce, using only already composted manures and materials is legally required. Instead, you can compost *underneath* your propagation tables. This allows the heat to keep the plants happy and gives the extra moisture from watering a useful place to go. At night, you can use hoops and drape floating row cover, plastic, or some other such material over the table to trap heat should the weather turn suddenly cold and the plants need additional protection.

Compost

A big way homesteaders save money is not just by growing their own food but by growing their own fertility. That fertility comes from compost. All the scraps of life that appear to have no value or life beyond the rubbish bin—coffee grounds and eggshells, banana peels and beef bones, cardboard and vegetable cuttings—well, they are a homesteader's future food. There are lots of easy and simple ways to get started composting; here is the one we originally used.

Pallet Composter
Compost needs a few things to thrive. Carbon. Food. Water. Air. A pallet composter helps ensure the fourth, because the slats provide natural air flow.

Material List
- ▶ 4 Pallets
- ▶ Some scrap 2×4s or similar wood
- ▶ Nails or screws
- ▶ Scrap metal
- ▶ Wire

For our pallet composter, we used a smaller pallet for the front. If you don't have access to a shorter or smaller pallet, you can cut down a full-sized pallet. A shorter front is necessary to make adding materials easier and to allow a pitched roof to keep rain and weather out. A compost pile needs protection from excessive rain, so don't locate yours under the side of a building with no gutters or in some other problematic place. Too much rain can harm the pile, causing nutrient runoff or drowning the compost and making the pile go anaerobic. This can make your compost smell bad, attract pests, and even contaminate your groundwater. So don't skip the roof. It is easy enough to make with some scrap sheet metal and wire. Allow it to overhang both the top and bottom a few inches.

If you think your pile will need additional aeration, or you just want your compost to finish faster, some scrap pieces of PVC or similar pipe with quarter-inch holes drilled all over them can be layered through the pile as well, sticking out the sides of the pallets for support. This

The Power of Pallets

Pallets used to be treated with a number of toxic chemicals—indicated by a CT on the pallet stamp—to protect them from insects and from moving dangerous insects like the Emerald Ash Borer around the world. While almost all modern pallets are now heat treated—indicated by an HT on the pallet stamp—many CT pallets are still in circulation. Also, an HT pallet may have had chemicals spilled on it that are not safe to have in or around your garden or animals. Given the number of pallets available, I stick with clean, HT-stamped pallets and encourage you to do the same. CT pallets are not suitable for your growing spaces and pose risks to your animals as well. Animals often bite and chew on wood and other such things, so I don't want my future meal consuming dangerous compounds that

will end up on my plate or that can sicken the animals or cause reproductive issues. So keep this pallet wisdom in mind when we come back to them in later sections.

A pallet should have a stamp allowing you to assess its appropriate uses.

provides additional air deeper into the pile, helping the material to compost more thoroughly and finish faster without needing manual turning.

When you start a new compost pile, make sure to layer the bottom with a *lot* of carbon—wood chips, sawdust, or similar material. Go for at least 12–24 inches. This helps prevent any runoff and the nutrients it carries with it from seeping into the ground and possibly contaminating groundwater or overloading the soil nearby, and, instead, saves them for your garden.

Portable Compost Tumbler

This was our big, early spring DIY project of 2017. We make a fairly large amount of vermicompost on the farm for both our use and for sale and barter. We also make an even greater amount of traditional compost, a small portion of which is used to make or amend growing spaces that need a very fine seed bed. Sifting compost by hand and picking out worms quickly became far too laborious, even for my son, who was paid fairly well for a nine-year-old. So, we decided we needed a compost tumbler to separate the finished compost from the worms and the unfinished stuff. We also use this sifter on our regular compost as well, where it allows us to collect a nice, fine finished material that is excellent for potting and plant propagation.

Unfortunately, all the ones we found for sale were expensive. Really, really expensive. Multiple thousands of dollars expensive, including used setups. Given I was doing only freelance work this year, a mid-multi-thousand-dollar expense on a barely multi-thousand-dollar enterprise didn't make much sense. The cost of a used unit was over twice our annual sales.

So my son and I decided to DIY an affordable compost tumbler. This project took us about six hours the first time, but we were taking it easy and talking through design options and scavenging materials from around the farm as we went. It was completely built from what was at hand, save the bolts, washers, and nuts. The second cage we built took only about an hour. If you hustle and have the tools and other stuff ready, this project should take half a day max if you are relatively handy.

Material List

- ► Scrap 2×4
- ► 50-gallon steel barrel
- ► Nuts, bolts, and washers
- ► Concrete reinforcing wire
- ► ¼ inch × ¼ inch hardware cloth. (Note, you can also do ½ inch × ½ inch, or $\frac{1}{16}$ × $\frac{1}{16}$. It just depends on the fineness of finished compost. For separating and saving worms, ¼ inch × ¼ inch is the largest size you want to use.)
- ► Four 2- or 3-inch wheel casters (fixed is best, but you can make it work with swivel style if that is all you can find)

Steps

1. Using a tape measure, marker, and string, mark the barrel in half around the middle.
2. Using an angle grinder, cut the barrel in half.
3. On the bottom end, butt out a hole large enough to dump a bucket or put a shovel full of compost through. This will be the *loading* end of the tumbler.
4. This is the trickiest part. Cut a section of CRW that is about 12–16 inches *longer* than the circumference of the barrel. Bend the CRW into a cage, similar to how it was turned into a cage for trellising, except with more overlap.

5. Test and see that the CRW fits *snugly* into both ends of the barrel. If so, secure it together using wire.
6. Now, wrap the CRW cage with the hardware cloth. The cage will be slightly longer than the hardware cloth, so center the cloth on the cage. Secure it in place using wire. Another option is to place the hardware cloth *inside* of the CRW cage. We think the latter is preferable but need more testing to be sure.

7. Place the hardware-cloth–wrapped cage back into the barrel ends.

8. Important step: It is time to secure the cage to the barrels. Make sure that the cage and barrel ends are square. Then, use a drill and appropriate metal-cutting drill bit to drill three holes in each barrel. Drill one at a time in a place where the bolt and washer will tighten down over a *corner* of the cage (where two pieces of wire come together), and then insert a bolt from outside the barrel inside; then add the washer and nut and tighten. As you continue to drill and add bolts, make sure that the cage and two barrel ends remain *as square as possible*. While it doesn't have to be perfectly square, the squarer the setup the smoother and more easily it will spin.

Our first frame was too narrow, causing compost to collect on the rails, and is up for redesign this year.

You now have the tumbler proper but need the frame and wheels that it will spin on. Our first frame design worked well overall, but the rails were too close in, and lots of compost tended to build up on them as we sifted. Our next design will try a new rail layout and wheel setup to address this and a few other problems. The first frame took us about 20 minutes to construct and was made solely from scrap and extra materials and supplies.

You want the long runners to be longer than the tumbler to make for easy transport. We use the overrun on both ends as carrying handles, and this makes it easy for two people to move the sifter.

Fixed casters are superior to ones that spin but may be harder to find.

Size the frame to sit on whatever you plan to sift into—for us, this is our garden cart. This way we can tumble until the cart is full, then set the sifter down and move the compost to where it is needed, saving ourselves some extra steps and work.

Super Simple Compost Sifter

Once you have finished compost, it may need some sifting before you apply it to your growing spaces or plant starts. Sometimes, this is because most, but not all, the material broke down. Sometimes, it is because some parts of your mix weren't sized right (like large chunks of wood chips and branches). Or sometimes it is because your plant starts or seed bed need a very fine mix, so the extra work of sifting the compost is an actual savings in better germination and plant growth.

Here is a super simple compost sifter design that we have used to make thousands of pounds of sifted soil and compost. With this setup,

 Brian Gallimore's Cage Composting

Another simple way to make compost comes from Brian Gallimore, a homesteader friend in Texas. It is especially designed for making leaf mold (leaf compost) and uses either CRW or welded wire formed into cages, similar to what I mentioned above for trellising, just much larger in diameter.

This type of compost maker works best with light, airy materials such as leaves and similar stuff. If you want to do heavier materials, you will need to reinforce the setup by using T-posts and two or three layers of wire to support the heavier materials' weight. Either way, it works well, providing lots of air to the pile, especially if you keep the diameter below eight or so feet. During excessive rain, you can cover the top of the cage with a tarp, cardboard, plywood, or similar material to protect the pile, depending on the gauge of wire you used to make the cage. Like with the pallet composter, you can also run pipes into the pile to provide better aeration instead of occasional turning if you are using heavier inputs.

Brian makes leaf mold compost in very large batches.

you want to size your frame to sit on your wheelbarrow, garden cart, or similar tool for moving loads of heavy stuff. This way, as we sift, once it is filled, we don't have to do extra work to move the sifted compost a second time and can take it straight to where we want it. Even though we have the tumbler, we still use this sifter often, especially for coarser mixes and materials.

Material List

- Scrap 2×4 or similar wood
- ½ inch × ½ inch hardware cloth (or smaller if you want a finer finished product)
- Screws

HOMESTEAD HOLLER
David Good's Surprisingly Good Homemade Potting Soil

Potting mix can be pretty pricey. My fellow homesteader David Good created a recipe years ago that uses all low-cost to free inputs. Ever wonder what to do with old, rotten wood that is no good for fires? This recipe puts it to good use. For this potting mix combine

- 1 part shredded, old rotten wood. We use a hatchet along the side of old rotten pieces of firewood that expired before they could be inflamed.
- 1 part aged cow, horse, or similar manure
- 1 part chicken house bedding/material

Sift the cow manure and chicken bedding. Then mix all three components together thoroughly. You should have an excellent basic potting mix that is appropriate for a wide variety of plants.

Speaking of sifting, you might wonder how do you sift this stuff? Take a look at how to make a simple sifter, on the previous page.

Credit: David Good

David's low-cost to free inputs make a great replacement for a basic store-bought potting mix.

Lay out the size of sifter you want, based on what it will sit on. Cut the hardware cloth and lumber to the right size to make the frame. Along one side, to secure the hardware cloth, you will double up the 2×4s to pin the cloth in place. If your sifter is very large, or you are using really heavy material, you may need to offer some cross supports to protect the wire from breaking under the strain. These can be wood or metal.

Once loaded, we run the smooth back side of a metal rake or shovel back and forth to sift. Or two people can shake the sifter instead. Both methods work well, depending on the amount and type of material.

We have screened thousands of pounds of soil and compost with this simple sifter setup. Note how it sits on and sifts right into the garden cart.

Extending the Seasons and Excluding Pests: Low Tunnels

For many homesteaders, the growing season is far too short, especially for our food needs. What if you could extend that growing season two or more months, while protecting plants from pests and other problems? You can, and you can do it for a low cost to start until you are ready for something larger. You can make low tunnels—small protected growing spaces generally just a few feet wide and high. Low tunnels only require two things—a hoop bender and EMT conduit. The conduit is available at any big box home supply store. The bender has to be ordered online. On occasion, you can find them used. I wouldn't hesitate to ask around in homesteading groups in my area to borrow one, though. It isn't a tool you need often, so it is best borrowed or shared instead of bought. One of the few things we purchased during our early years was a hoop bender, and it mostly spends its days collecting dust or finding its way into the hands of friends and other homesteaders in

Low tunnels let you protect plants from pests while expanding your growing season.

our area. I don't think I have bent a new hoop in half a decade, so I am glad to see it get use.

To cover the hoops, you have a few options. Floating row cover is great and is all some growers will need year round, especially if you are far enough south or just protecting frost- and freeze-hardy crops. In mid-latitude lands, floating row cover works for most of the year, but for late fall and through winter, switching over to greenhouse-grade plastic is important for protecting and preserving your plants.

The benders come in various sizes, so make sure you have a good idea what size of growing spaces you are going to stick with before buying one. You don't want to bend a bunch of pipe only to realize it won't work with your growing space sizes later on. We went with six-foot-wide beds, so we purchased a six-foot bender.

Multi-Use Makes a Big Difference

EMT hoops can serve a lot of other purposes beyond low tunnels. We have used them to make a small chicken brooder/shelter, a hothouse for improving germination of certain seeds and protecting young, heat-

Two pieces of slab wood, a few EMT brackets, and a couple of hoops can make a portable hot house/mini greenhouse. Once the plastic is added, this is easy to move anywhere it is needed.

loving starts, a portable plant protector placed on rails, and so much else. Whenever possible, try to find unique ways to expand the uses of a piece of infrastructure.

Earlier, I mentioned how even fencing has not protected our melons from hungry raccoons and possums invading our garden. But floating row cover has! We have to pull it back every so often to allow for pollination, but once the vines are well laden, we leave it on and keep those mean paws from making a meal of every last melon.

Simple Garden Gate

You have a lovely garden, enclosed in some kind of fencing, full of plants growing up and out, fed by compost from the scraps of your own kitchen and homestead. How do you get in and out of it?

Scrap wood and metal tend to build up, especially when you're first establishing your farm or homestead. Instead of purchasing expensive gates and doors, you can make simple yet attractive ones out of these spare pieces with just a few minutes of work.

Some scrap wood and wire can make simple yet fitting gates for growing spaces.

Material List

- ► Lumber for gate—I used rough cut 1.5×4s in cedar, but many kinds will work and last for many years
- ► Leftover fencing—I used welded wire with 2×4-inch holes, as we are keeping out rabbits and chickens
- ► Screws and metal staples

Steps

1. Cut the wood to the proper dimensions for the door—about 3–4 inches wider than the opening for the width and to height. I cut mine to the same height as the wire, so 48 inches.
2. Lay the door out square.
3. Cut the wire to the proper dimensions (just slightly less than the door's full dimensions).
4. The pieces of wood will help secure the wire in place. Remove the two higher pieces of wood. Lay the wire onto the remaining two squares. Place back the other two pieces of wood. Check that everything is still square. Drill two screws into each corner at a 45 degree offset from one another
5. To keep anything from pushing through the wire, use metal staples to secure the wire to the wood along the four sides.

6. To attach the gate to the fence, we used 12-inch-long pieces of leftover fence wire and lightly tied it in place.

Plastic Baggies and Pantyhose

A number of folks have suggested some other great garden tips. One, for instance, involves using plastic lunch baggies to protect fruits from pests. Another person talked about how you can use pantyhose to support trellising heavier squashes and melons to grow them up without breaking the vines with the weight of the fruits.

Offsetting screws keeps things square, and also applies to the simple sifter corners in the previous project.

While these are ingenious and creative ways to solve problems, I would ask if they aren't creating bigger problems in the long term. Plastic baggies and pantyhose will still be around in landfills and other such places five hundred to a thousand years from now. Creating relatively permanent pollution may be cheap for us, but what about the cost to our grandkids and great-grandkids?

I don't feel comfortable producing single-season waste like that for the sake of convenience on my homestead. Now, I still do use plastic, like on my high tunnel. But that plastic gets many, many years of use, and then is repurposed for weed control, low tunnels, wood protection, and small structures. Only then does it finally get recycled. I suppose if someone was super careful, they could get a few years use of out plastic baggies to protect fruit and pantyhose to support trellised stuff, and if you do use those approaches, I hope you will do as much as possible to minimize their environmental impact and maximize their usefulness over many years. Or maybe a used pantyhose recycling program is needed in some parts of the country. (Is this even sanitary? Hmm…)

Really, this is about more than plastic baggies and pantyhose. It is about the bigger choices we make not just as consumers but as co-stewards and communities. So as you build your homestead and try to do so affordably, don't cut corners only to pass the costs on to others who aren't even there yet to protest. Try to minimize creating waste streams that will long outlast you and your kin.

Don't Underestimate the Danger of Drift

When looking for a farm or homestead, there are a lot of things to consider—too many to list here (probably worth their own book). One that many current and existing homesteaders overlook is the risk of living next to industrial agriculture or a place that may be prone to herbicide overspray and drift. Not only does this put your well and other water supplies in peril but both your own agricultural efforts and animals as well. Overspray and drift, the movement of herbicides, insecticides, and fungicides beyond the point of application to adjoining properties, is a problem that is all too common. In any given year, hundreds of my friends and acquaintances are negatively impacted by it.

In 2017, millions of acres of land in the United States were hit with this drift. By conservative estimates, a hundred thousand properties or more were affected. A few of my friends lost their entire farming operations because of it. If your property is downwind or adjacent to a row cropper or other operation that sprays chemicals, be warned and do whatever you can to minimize the possible damage. This is one of those places where an ounce of prevention is worth a pound of cure. If you haven't purchased a homestead yet, look for a place that is somewhat safe from herbicide drift, if at all possible before you buy. Then, put some resources into building buffer zones of multi-layered and -heighted trees, shrubs, and other plants to create a line of protection if needed or even just in case things change in the future. We are fortunate that our 35 acres sit surrounded by thick woods on all sides, with no row cropping or similar operations close by. Yet we always keep an eye on adjoining land that goes up for sale and what it could mean to our property and plans.

If you do get hit, don't count on getting it sorted out in court. Most drift cases never get compensated, and if you do get compensation, it is often at best a fraction of the dollar value of the damages. It isn't right, but it is what it currently is, so do your part to protect yourself and your place.

Turn a Trampoline into Almost Anything

Trampolines—those once beloved, now "we will cancel your home insurance if you have one" wonders of childhood. With a life cycle of about three to five years, trampoline frames pile up faster than cars on blocks in rural Kentucky. Used trampolines dot the landscapes of both city and countryside. Most, after a few years of enjoyment, go straight to the dump. On occasion, they end up as recycled metal. The mat, springs, and safety netting usually give out long before the tubing/structure itself. The high-grade steel tubing often ends up in landfills, when instead it can be turned into all sorts of useful farm creations for storage, animal shelter, or small greenhouses.

This section will first outline the basic conversion. Then, we will talk about different directions you can go with the basic frame.

If you want a greenhouse but are not ready to go big, a trampoline version is a low-cost way to try one out. Other than the trampoline frame, some self-tapping metal screws, and the cross pipe, the only other supplies you will need are some lumber and the greenhouse plastic. The plastic will be your biggest expense, along with the fence-top rail or EMT conduit—if you can find some used but still serviceable stuff from another greenhouse grower, it could save you a hundred dollars or more. The same goes for fence rail or pipe—if you can find some free, it will save a fair bit on the total cost.

Basic Conversion

Material List
- Good-condition trampoline frame
- EMT conduit or fence-top bar railing—four to six pieces. Conduit is less expensive but not nearly as strong as fence rail. You will need to measure the trampoline frame holes to ensure you match the cross rail pipe size to the trampoline leg holes.
- Self-tapping metal screws
- Large, flat work area

Steps
1. Disassemble the trampoline frame.
2. Lay out two arches from the trampoline's circular frame on *flat* ground
3. Using the self-tapping metal screws, screw the frame together with two or three metal screws per joint. For two screws, they should be at around 4:30 and 7:30 on the tubing on what will be the inside of the frame.
4. With help, insert the cross pipe into the trampoline frame feet.
5. Once the cross bars are in place (use a rubber mallet to drive the cross bars and hoops tightly together if needed) and the design is square, use two metal screws per joint to secure the cross bars. Again, these should be at 4:30 and 7:30 on the piping on the inside of the frame.

Make sure the outside surface of the frame is smooth, since any sharp edges can tear whatever you cover it with.

Important Note: All of the metal screw heads should be on the *inside* of the frame. Also, the screws should be short enough that none pass through the entire metal tube, so we use 1½- or 2-inch size screws. The outside of the frame should have no protrusions, sharp edges, or other such things that will either damage or reduce the life of the covering.

A Few More Construction Tips

You should now have a 10-foot-long, more or less 15-foot-wide hoop house frame that, if you laid it out properly, is relatively square and level. But what do you do with it next?

No matter what you turn it into next, you need to make sure it is properly secured. One option is to drive the four legs into the ground about 18 inches deep. You lose a fair bit of height this way, so I suggest you save the original trampoline legs (about 24 inches or so in height; if they are too tall, you can

cut them down with an angle grinder or a mitre saw with a metal blade). Then, I drive these into the ground as ground posts and set the frame into the posts, once again putting two screws per joint to secure the sections together. If you are in an exceptionally windy area, I would upgrade from the two screws to a single lag bolt at each joint, which goes through the entire pipe on both sides. Make sure the smooth, rounded head side of the lag bolt sits on the *outside* of the frame.

Where You Can Go from Here

If you build your trampoline greenhouse according to these instructions, you can easily remove the plastic once it gets warm enough and turn it into lovely arched trellising. Or, if you decide you don't like it as a greenhouse, you can wrap the frame in a number of different types of wire—welder, concrete reinforcing, and some others—and grow all over and under it. In winter, when the growing season is through, you can toss a cover on and use it for storage.

Trellised Growing Arch

One option is to take concrete reinforcing wire or another type of wire and wrap it over the frame to make a large, arched trellis space for growing. The frame will provide a great deal of support for even heavier vegetables and plants to grow, while creating a large, shady microclimate underneath for ground crops that needs less intense light and heat as spring turns to summer and temperatures rise.

Small Greenhouse/Hothouse

This is one of my favorite ways to use these small structures. The size makes providing heat for plant starts in spring pretty easy without breaking the bank, including through something as simple as adding a lot of thermal mass (50 or so one-gallon dark-colored plastic drums filled with water).

If you go the greenhouse route, there are a few choices you will need to make for how you finish out the structure. Will you frame out the end walls, making it more permanent, or just let the plastic drape over the building, securing it with bales of straw or similar weighty objects? Framing it out will make it far less portable, depending on how you frame it, but will make it far more wind resistant and increase the life of the greenhouse plastic, which is the largest expense for this structure.

Either way, if you decide to cover with greenhouse plastic, you will need to add wood baseboard on both sidewalls at the base of the structure. Then, you can either use wiggle wire and C channel to secure the plastic, or you can sandwich the plastic between the baseboard and another piece of wood. (I would recommend using a 1×4 with screws.)

The baseboard would be attached near the joint that anchors the structure into the ground, so the lag bolts mentioned above would need to be long enough to go through not only the metal tubing, but also the wood baseboard. For the baseboard, I recommend using 2×6 cedar. You can use treated lumber, but then will need to not grow in the ground within 6–12 inches on either side of where treated wood comes into ground contact.

Animal Shelter

For the first few years on the farm, we didn't have a good shelter for larger animals in winter. So we used the trampoline frame to create a winter-time loafing area for our cows and provide them some relief from wind and rain. We covered the structure with a used billboard tarp. Some people use regular tarps, greenhouse plastic, old silage tarps, or another suitable covering. Like with turning the frame into a greenhouse, the covering needs to be securely attached or anchored around the frame.

To do this, we used a large number of concrete blocks. You could also use straw bales or any other heavy objects. The covering will need to extend two feet or so along the ground out from either side of the structure to provide ample space to anchor it in place. The tighter it is, the less likely to blow away.

We left the open side of the structure facing south and east, a direction we rarely get wind or weather from. This left the north

This provided a covered space for our cows through winter that they and the chickens thoroughly enjoyed.

and west opening to deal with. While the billboard tarp provided some coverage along this wall, we used a combination of sheet metal and straw bales to provide a strong wind break. As winter became spring and the weather warmed, we used some of the straw bales to rebed the cows until they could be let loose back out on pasture again.

Equipment, Hay, or Other Storage

If you are short on protected storage, this structure is a simple way to create some additional space to keep weather off of equipment or animal feed and bedding. When I was first considering this conversion, the first way I saw it used by another homesteader was to protect their small tractor and some other equipment.

So Many Other Options

You can use this basic design in a lot of different ways—movable small-scale greenhouse, chicken brooder for pastured poultry, hothouse for plant starts. At a total cost of around $50 or less (not including covering material), the cost-to-usefulness ratio is through the roof, and your options are endless.

Trampoline Chicken Run

Another great way to repurpose a trampoline involves keeping it mostly as is but cladding it in protective wire to use as a chicken run in conjunction with a chicken coop or house for night-time safety.

[2]

For Your Animals

An animal a day keeps the doctors away.
— John Moody

If anything has defined our farm, it is the unending series of chicken houses we affectionately named "Fort Clucks" over the years. We have always had chickens. Probably, we always will have chickens. We also have a neighbor with a 5,000-acre wildlife preserve who rehabilitates predator animals and releases them into the wild faster than a preschool multiplies and releases the stomach bugs across its students during cold and flu season. So we have always needed Fort Clucks. Typical chicken protection protocols just won't do.

We built some versions from straw bales. Some from an old round bale feeder. Even today, the back of our barn resembles some dystopian building straight out of The Walking Dead, with used tire sale signs and wire reinforcing key entry points. Some did better than others. We are on Fort Clucks version 5.3 this year. None cost us more than a hundred or so dollars, and, really, they all cost almost nothing since the materials were all repurposed for other projects when the latest fortress finally fell.

Like people, animals need food, water, and shelter. Depending on the animal, it needs not just protection from wind and rain but from predators big and small—flying, squeezing, and digging. Animals generally represent one of the largest investments on the homestead. A laying hen or meat bird represents around 15 to 20 dollars each in value.

You can make a simple chicken tractor, suitable for a small number of birds, from scraps of metal and wood—like our first portable Fort Clucks.

Larger animals like pigs and lambs go into the many hundreds. A dairy cow may run you low four digits. Ensuring they are well provided for and protected should be a high priority. But it doesn't have to come with a high price tag. Unfortunately, a lot of people only learn how to raise animals affordably the hard way.

Fencing

It is very important that you keep your animals contained where you want them. I might add, because a few folks don't seem to realize this, you want them only on *your* property, unless you are leasing or otherwise have access to additional land not your own. Remember: "Good fences make good neighbors." If you are not keeping your animals contained, you are risking all sorts of problems, including legal. Escaped animals that cause accidents or damage to other property or people will not merely put you in the dog house with neighbors—they could put you in the courtroom or jailhouse.

Once, I paid someone down the road a hundred dollars for the damage an escaped cow did to her garden. A few other times I have had to chase cows across multiple acres of adjoining roads and pastures. A few calls from neighbors was all it took to make us realize that if the animal wasn't going to willingly stay put, and nothing we did was going to change its mind, then it needed to go.

Also note, while some in the country think it is okay to let dogs roam free, if your dogs hurt or injure people or livestock, you are going to (and should) pay. So, take fencing and property lines seriously, and view respecting and maintaining them as an important investment both in your personal sanity and community standing.

 ## The Most Expensive Cheap Eggs You Ever Had

At least in my area, lots of people get into backyard chicken keeping. It is such a popular hobby that for a few years my daughter made great money raising laying hens from day-old chicks to pullets and then selling them off at a tidy profit. Most people do it to save money, since good-quality eggs are often considered expensive. Some close friends of ours learned that those good eggs are actually under priced.

They did what most urban homesteaders do. Build some raised beds. Catch water from the roof. Grow herbs and other things in pots. Build a chicken coop. Buy some chickens. Many people don't realize that chickens don't lay all the time. Actually, they don't lay a lot of the time, especially if you don't know how to keep them. Even if you do, between winter slowdown to molting, there is a fair stretch of the year that chickens don't earn their keep.

For my friends' tiny backyard flock of five layers, they were going through 20 pounds of feed per week! Eventually, they ran the numbers and realized that their low-cost eggs were coming in at over seven dollars a dozen, and that didn't include labor or the cost of the coop and its upkeep. So they got rid of the chickens and started buying eggs again to save money.

Some endeavors will save you money, but many, especially in the short term, may not. My friends learned this the hard way and in trying to save actually ended up spending almost twice as much instead. Animals are best viewed and understood as a *long-term investment*. It will generally take time before they fully pay for themselves and the infrastructure you need to establish to keep them. There is a lot you can do to reduce that return-on-investment timeline. All the same, it still takes time before animals make sense and cents.

Credit: Randy Buchler

Randy Buchler and many other farmers use pallets for fencing and runs around their homesteads.

Keeping Them In (and Out): Pallet Fences and Gates

If you need a corral or small paddock for your farm animals, a pallet fence may be an affordable option. If your animals are aggressive diggers, like some hogs, you will need to run a hot wire along the inside bottom of the fence about six inches high or so to keep them from going under, but otherwise, most animals take to pallet fences fairly well. They are also especially good at providing shady spots and protection from wind. Just note that if the animals spend a lot of time in one area or side of such a fence, you will want to add a lot more bedding/carbon to these places—wood chips, sawdust, straw, or whatever is at hand—so that manure buildup or other problems like compaction are prevented.

Cattle-Panel Corrals

For certain animals that we cannot let freely roam our property, such as our pigs, we use cattle panels and T-posts to create enclosures appropriate for their kind. I realized that these panels, at 18 feet in length, allow you to put up long stretches of fence quickly but also come with some significant drawbacks. So I have made a key change to how we use cattle panels. Instead of leaving the 18-foot panels intact, I cut many of them in half. First, this makes it so I can actually get them home from the store safely—few homesteaders have a truck or trailer that can handle materials of such length. It also makes them far easier to handle and place more tightly along the fence bottom to prevent escapes, while also providing multiple spots where we can easily get material or animals in and out of their enclosures. Also, these shorter segments are much easier to handle solo.

We often keep pigs on our compost field—a quarter-acre area set aside for making tremendously large amounts of compost—for nine or so months of the year (with a 90-day period after the pigs have left for the material to finish composting and for manure and other material to safely break down). The pigs turn our compost and the compost provides free food for the pigs. Once fall sets in it also provides a tremendous amount of free warmth for them as well, reducing feed costs and improving their weight gain since they don't have to waste as many calories keeping themselves warm—the compost field does it for free! Because we are adding to

Some wire plus a pallet provides a perfect access gate to move compost from the compost field to the growing spaces without having to go all the way around the entire corral.

the field weekly, along with accepting loads of wood chips and other materials, the cattle panels have to remain semi-movable so that we or others can drop more material into the field. The nine-foot panels are the perfect size, though we occasionally need to slide two out of the way to make a wider slot for a large drop. The other benefit of the cattle panels is that they contain pigs but allow our chickens access to larger areas since they can easily get through the holes. Pallets also make great, quick gates. I especially like using them with cattle panels to provide easy access points to move our garden cart, round bales, and similarly sized stuff in and out of animal areas.

Scrap Wire and Metal Saves the Day

The weakest part of an animal enclosure or house is generally under the ground, not above. Most predators are excellent diggers and in just a few minutes can burrow under the walls to get at the animals inside. How do you stop them? The simplest method is to bury wire around the outside of the structure or enclosure, especially poultry housing. Leftover scraps from fencing and other projects are perfect for deterring predators, and this is one of the few possible reuses, other than recycling, for such scrap. Predators paws don't enjoy the pain of digging into metal wire. They are adapted for dirt and similar debris, not wire. The metal almost immediately gets them to move elsewhere in their search for a meal.

We like to have the metal come out a minimum of 12 inches from the base of the structure. If it is too short, the predator will just dig a bit farther out and go under both the wire and the wall. The wire needs

to be secured in place, either by dirt or some other heavy material. If it isn't, an animal may just push it up against the wall and then start digging where it used to sit on the ground. Since burying wire around our new chicken house, we have had zero night-time losses to predators. Almost every night when I go check on a few things at the barn and pass the coop I watch a number of predators scurry away who have been seeking entry or loitering close by—but still no losses. It is a simple and effective solution to night-time snacking in our neck of the woods.

Keeping Them Fed and Watered

One of the first ways we got into low-cost homesteading was through a friend who worked for an independent grocery store in Louisville. He was in charge of the produce procurement for all three locations, and they had, like any store, some loss every week. The organic juice bar was also located in the store, and also produced a lot of waste that had to be disposed of in some way. So he asked me if I would want to pick up the material.

 ## A Warning about Chicken Wire

Chicken wire has one purpose: to keep chickens in (or out). Against a wide variety of predators, its protective value is debated at best. I can no longer count all the stories shared by my farming and homesteading friends of predators passing through chicken wire as if it didn't exist, to access the defenseless animals inside. Persistent predators, especially medium-sized dogs and other more powerful animals, can easily push and claw through chicken wire. Dogs are a real problem for urban and suburban homesteaders, and they make quick work of chicken wire if they are of sufficient size, determination, and disposition. Chicken wire is one of those "cheap but expensive" things in life.

Given the cost both in terms of time and money, I personally suggest you don't risk your animals' lives by trusting them to chicken wire. If you use it for the sides of animal housing or tractors, it will need to be reinforced by wood or metal cross pieces to provide strength and resistance against aggressive predators. Otherwise, upgrade to welded wire or hardware cloth with appropriately sized holes (no more than 2×4 inch). Chicken wire is great to reinforce slatted wood and small open spaces, such as the sides of pallets and structures made from them. But otherwise, chicken wire is for the birds.

We use chicken wire to allow greater air flow and circulation in animal shelters but only to cover smaller, very secure spaces.

As this second picture shows, chicken wire is only a small deterrent to a persistent predator.

There is a tremendous amount of free food, of edible quality for both humans and animals, that is tossed out daily around the United States and can become a great source of supplemental feed.

Pretty soon I was dropping in twice a week and collecting over 2,000 pounds of pitch produce—items that are out of date, of poor quality, or otherwise unsellable but still relatively edible—and coffee grounds from a handful of locations across town, providing a literal ton of free feed for our animals, worms, and soil. Since we were already going to town twice a week for various commitments, the extra 30 to 45

A small coffee shop produces five or so gallons of coffee grounds *per day*. This makes a great soil amendment, compost addition, worm food, and so much else.

minutes was more than covered by the value of all the compost. It also allowed us to treat our trips as business and thus write off the mileage as a business expense. Even without the tax benefit, it was still a worthwhile addition to our schedule.

We are now at a point where we can generate almost all the fertility our farm needs on site, and anything else we bring can be used to improve pastures or made into sellable compost. But when we were first starting out, access to such free resources allowed us to move our soil and animals along far more quickly and affordably than any other options or methods.

Pitch Produce

Animal feed is often the biggest single expense on a farm, especially for chickens and hogs. Traditionally, these animals had little access to bought grain, as grain was expensive. So they had to scratch, peck, and recycle whatever happened to be available beyond a small bit of grain that farmers gave them. Pigs cleaned up fields after harvests, such as peanuts and sweet potatoes, and then moved into fall surviving on fallen nuts and fruit. Chickens kept bug populations in check while producing excellent eggs. When winter came the eggs went away. Orchards were cleaned by turkeys and small-sized breeds of pigs passing through, or by the dropped fruit otherwise making its way to the appropriate animals.

While many farms no longer have these resources, large cities often provide ample calorie streams that can be redirected back as animal feed. Estimates show that as much as one third to one half of all food produced in the US goes uneaten. A large portion of that ends up in landfills or otherwise unceremoniously discarded. Entire orchards of apples and peaches get dumped on the ground. Entire shipments of bananas that ripened a bit too fast go to the dump. Dozens of gallons of arbitrary date-stamped day-old milk are tossed in the waste bin. The amount of stuff is endless. Its value as both animal feed and compost to improve soil is unimaginable.

If you already have to head to town, why not make the most of the trip and snag food for your animals and fertility for your garden at the same time?

A Warning about Waste Streams as Animal Feed

It is very important to note that different states have different rules and regulations covering what you can and cannot feed to different species of animals. So, before you go all in collecting brewer's waste or juice bar leftovers, make sure you know what you are allowed to feed to what animals and any other rules and regs you will need to follow. Sometimes, this applies only if you are selling your animals or their meat, but sometimes, state laws do not differentiate between what is just for you and what is for market.

You can check with your local USDA extension office as a good starting place to find out what rules you may face. The EPA and Harvard also have excellent resources. Two online resources you can use to check your state rules are

- epa.gov/sustainable-management-food/reduce-wasted-food -feeding-animals
- chlpi.org//wp-content/uploads/2013/12/Leftovers-for-Livestock _A-Legal-Guide_August-2016.pdf

A worm bin can produce the perfect plant and animal food at the same time.

Other Ways to Cut Down on Feed Costs

Beyond seeking out pitch produce and other food waste streams, there are a lot of other ways to save on animal feed. Especially for poultry, raising worms (either composting or meal worms) is top of the list. Some of my friends also raise crickets seasonally as a food source for their fowl. Both do well on nothing but waste streams, turning kitchen scraps, coffee grounds, animal manures, leaves, cardboard, and other such marginal stuff into valuable compost and low-cost or free poultry feed at the same time. You can learn more about our IBC tote worm compost bins on page 106.

Pig Barrel Waterer and Feeders

Many folks water their animals by filling open dishes. While easy to do, it also has a lot of drawbacks and dangers. Water in open containers is often fouled by our animals, especially chickens and pigs, who get mud, manure, and other muck in the water. Pigs are the worst. Instead of just putting all the above in the water, they will put themselves into it, mud, manure, and all! A lot of water is spilled from open containers, which, if you are in a water-stressed area or have limited water supplies, isn't wise. Spilled water in the same spot can start to damage and degrade the ground around your watering spots, especially if you are already dealing with saturated soil. Standing water is an invitation to all sorts of pests to multiply, especially mosquitoes and other unhappy insects that afflict us homesteaders.

Clean water is key to animal (and human) health. Providing water that is unsullied by animal actions is an important part of good animal care. Nipple waterers—made specifically for each animal breed, since each has slightly different drinking motions—are a low–cost way to ensure clean water. You can plumb them into buckets, plastic or metal pipe, or barrels and totes to give the animals access to small amounts of water on demand at any time while protecting the reservoir from contamination.

For small sets of pigs, a 30-gallon or 50-gallon poly barrel makes an excellent waterer. The reason to keep the barrel up on a pallet is that otherwise the pigs will turn it into a wallow in no time. This will eventually cause the barrel to

Do you see the chickens' water dish? Neither do I.

These can be plumbed into a bucket or into small pipes to give chickens constant access to clean water during above freezing temperatures.

A 30- or 50-gallon poly barrel plus a few low-cost pig nipples makes an excellent water option for a small number of pigs.

topple over into the mud and muck and create a real mess for you. (Trust me, I know!) Because pigs are powerful and inquisitive animals, we also keep them from tipping over the barrel by using a few spare T-posts to pin it in place on top of the pallet. While a full barrel weighs over 350 pounds, pigs have no trouble knocking it over and turning it into a plaything. Other friends located their barrel outside the fence, with just the nipple reaching through to give the pigs access while solving this problem of their knocking it over. A few farmers also plumbed pipes off the barrel into their pig enclosures as another solution to keep the barrels from becoming pig toys.

A pig waterer takes just a few minutes to make. All you need is a drill, two or three pig nipple waterers, plumber's tape or putty, and the right size drill bit for the size of pig nipples you purchased. The product or the seller should make it clear what size hole the nipples fit into and what size drill bit you will need to install them.

It is best to mount the nipples about 12 to 16 inches from the bottom of the barrel—this does mean that you lose 10–20 percent of the barrel water volume, but that means you always have some ballast in the bottom to keep the barrel in place. We use two nipples per 50-gallon barrel and find that more than sufficient for four or so pigs. This makes watering a twice a week or so chore instead of another daily duty, except during the hottest parts of the summer, when every-other-day or daily refilling sometimes becomes necessary.

If you have open air cattle troughs for water, you can keep some goldfish, guppies, or similar aquatic species in the tanks to eat any mosquito and other bugs that come for a visit, or at least, the eggs they leave behind.

If you toss feed on the ground, be sure it is during non-rainy weather and only enough for the animals to consume quickly, otherwise much of it may be lost to rain, rodents, and other pests.

This style of barrel is often less expensive than many others because it is also less versatile.

Pig Trough

Dumping feed on the ground isn't generally good for the feed or the animals. A lot of feed is lost unless other animals come along to clean it up. Sometimes, the cleanup crew that comes along are not animals you want but mice, rodents, and other pests and predators.

For pigs and chickens, a feeding trough is a simple way to save feed. For the chickens, an old piece of PVC pipe cut in two or a simple wooden trough works wonders. For pigs, it is as simple as an old, food-grade barrel cut in two. I especially like to use the bunged liquid barrels since they have far fewer uses than regular, lidded ones.

Once halved, these barrels make excellent feed troughs, small-scale worm compost bins, and more.

Pigs often will move their feeders, sometimes quite far. We secure the troughs to a T-post by drilling small holes and tying the trough to the fence with wire run through them.

Tip: pigs love to pull their feeders around. So when we made troughs, we drilled small holes that allowed us to tie them to a T-post and the fence with some wire so that the pigs couldn't drag them all over the place and turn them into toys.

Kid-Sized Infrastructure

Buckets are an indispensable part of homesteading and farming. Most people have large piles of five-gallon buckets for moving water, feed, and who knows what else. Even full-sized adults may struggle to move fully loaded five-gallon buckets of water and feed. What about trying to get your kids involved? Such sizes make it highly unpleasant if not impossible for them to contribute.

 ## Carlos' Small-Animal Hay Feeder

Feeding out square bales can be tricky with small animals. You want to maximize access while minimizing loss. Carlos' small-animal hay feeder is a perfect solution. Low-cost, portable, and easy to build, this feeder is great for a wide range of animals—sheep, goats, alpacas, and the like. If you use it with larger or more aggressive breeds of cows, be careful, as they can easily knock it over, especially if overcrowded.

Credit: Carlos Cunha

When sourcing buckets, make sure to get some smaller sizes that allow your kids to succeed in serving as a help around the homestead. Two- and three-gallon buckets are the perfect size for kids 12 and under (and for a lot of adults, as well). For kids under seven, one-gallon buckets, or even milk jugs with the tops cut off are good options that allow them to get involved. Smaller buckets are not hard to find—ice cream shops, bakeries, and many other businesses produce plenty in these sizes.

When we do work around the farm, I often also have extra tools that I let the younger kids play with

Larger buckets can weigh twice as much as smaller ones, making tasks too hard for kids (and adults) to help. Make sure to keep some smaller buckets on hand for both big and little helpers.

When the Kids Commandeer

Don't be surprised if your low-cost projects get commandeered by the kids for other purposes. My trampoline animal shelter has spent the last two years as our kids' "practice field," a makeshift playground that they swing, climb, and play all sorts of games on when weather permits. Last fall, they took a spare pig trough and turned it into their first "boat." Of course, it was for sending their little brother out into the wild waters of our pond—the older siblings claimed they were too heavy so they volunteered their younger brother for the christening voyage. I keep asking if they took out life insurance first but have received no response.

If you want your kids to enjoy your homesteading adventure, they need these opportunities to take ownership and create memories while also learning life skills that will benefit them wherever they go in later life. Let them make use of stuff. Let them make stuff. Let the homestead be as much theirs as yours, so that you can say together one day, "This is ours."

After our cows went off to the freezer pasture, the kids took their old shelter and turned it into hours of endless play.

in appropriate spots as well. Same in the house—an extra broom is an inviting distraction to a one-year-old who wants to do what she sees her older siblings getting into instead of her getting into trouble or making new messes while unattended. Make homesteading attractive and accessible and your kids will want to do it with you from day one.

Keeping Animals Warm and Dry

Barns are expensive. For many animals, barns are also a less-than-ideal environment for their well-being. Air quality, light, and moisture levels in a barn generally are not the best. Also, such secure, covered storage space is usually at a premium on the homestead, so if we can make more appropriate shelters for our animals *and* free up barn space for storage of other valuable stuff, we have a win/win situation. But if not a barn, what will you build for your animals? The options are pretty limitless, and you can appropriately house any species of animal affordably, both young and old.

Simple Pig Shelter

The first year we had pigs, we had no shelter for them when weather became unfriendly. With some scrap lumber and extra sheet metal, we made a simple roofed shed structure to project them from wind and weather. It took us about half an hour and a buck or so in screws to make our first set of pigs much happier. Instead of supporting the high side with more wood and framing, we used the fence itself as part of the structure and allowed the roof to sit on the fence. It did require us to add a few additional T-posts on the fence side to support the extra weight.

This took us about 30 minutes to put together and also provided a sheltered spot to feed the pigs in bad weather.

 **HOMESTEAD HOLLER** Sean Metzel's Chicken Tunnel

Far north, winter is hard on animals of all sorts, two- and four-legged. My friend Sean Mitzel homesteads in the northern regions of Idaho, where the winters are long, cold, dark, and incredibly snowy. Such weather makes keeping animals extremely difficult and expensive. Water freezes. Extra food is required to keep up the animals' condition. But not for Sean, since he found a solution that keeps his animals warm and dry while providing food and unfrozen water all season long—his chicken tunnel.

Sean's chicken tunnel is dual purpose—winter housing that helps the birds lay better by staying warm and dry and provides free food from the deep-bedding growing space for their garden, spring through early fall.

Fort Clucks 5.3

Early in the spring, one of our neighbors purchased a riding lawn-mower. A big truck somehow managed to back itself up our gravel road and drop the thing at the top of his perilous driveway. As he unpackaged the massive machine, he ended up with a stack of heavy-duty pallets just sitting at the top of the drive. Day after day, as I went to and fro, those pallets got my eye and my imagination. We were getting ready to get new chickens for the farm after a series of unfortunate events had depleted our previous flock, and a new Fort Clucks was in order.

I asked him if he needed the pallets dealt with and he said, "Please!" Fort Clucks 5.3 was born. Between scrap lumber and this massive pallet set, we were able to build Fort Clucks in about six hours for $50 or less (see next page).

This design could easily be adapted for a kid's fort or playhouse (my kids pushed hard for this option!), a storage or bike shed, or some other purpose. If you build a coop, it is best for it to open to the east, both to minimize incoming weather and allow the early morning sun to hit the birds as soon as possible each day.

Brooding, Roosting, and Bedding

Animals need a lot more than just a home. Some, like chickens, need different quarters for the first few weeks of life. Many types of poultry like perches or roosts inside their special spaces. All need lots of bedding to ensure they stay adequately healthy and to prevent the loss of valuable nutrients in their urine and manure. So let's look at options and ways to save in each of these areas.

Simple Chicken Brooders

Everyone loves chickens. I don't know if I have ever met a homesteader who didn't have chickens. Most folks buy day-old or few-week-old chicks and raise them to laying or butchering age. But baby chicks are fragile little folks. For the first few weeks of their lives, they need a stable temperature and snuggly environment to keep them safe and out of trouble.

When I saw the pallet sitting at the corner of the road, I knew exactly what I wanted to do with such a great resource.

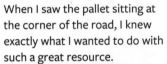

 Veldcamp's Very Nice Mobile Chicken House

If you can get your hands on old metal frames, say from farm wagons, trailers, or similar things, the sky's the limit for what you can build. Old school buses are another great option for storage or animal shelter. Smaller frames are a special favorite of people wanting to build portable chicken coops, which have many advantages over stationary.

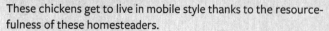

These chickens get to live in mobile style thanks to the resourcefulness of these homesteaders.

Jordan's Green's Magnificent Multi-Use Animal Shelter

Jordan Green is a farmer friend who produces a lot of poultry and pigs. Since his operation is pasture based, he wanted to keep his animals out on pasture as much as possible.

This is his material list and basic design for an animal shelter that works for almost any small animal, including smaller breeds of cows or calves from most any breed. The shelter's durability and flexibility combined with its relatively low cost make it a great way to give animals both shelter and pasture at the same time. It is made to be pulled by a truck, tractor, or even suitable riding lawn mower.

Jordan uses these shelters for farrowing sows, but they would serve various types of medium- to small-sized livestock well.

Lumber (all pressure treated)
2   2" × 12"—10 ft.
1   4" × 4"—12 ft.
3   2" × 6"—8 ft.
11 2" × 4"—8 ft

Plywood
1½ sheets of ¾" or ½" plywood

Metal
2 sheets of 36" × 96" corrugated
    metal

Hardware
2   4" × ½" eye bolt
2   5" × ⅝" carriage bolt
16 6" timber lags (lag screws)
40 5/4" sheet metal screws
50 2½" decking screws
50 3½" decking screws
    4 ft chain
    2 chain clip

Depending on the number of chickens, there are a lot of options for reusable, repurposable stuff. For instance, the same hundred-gallon Rubbermaid water troughs that many homesteaders use for aquaponics and other purposes also make great temporary brooders for 5–15 chicks. Many times over the years we have used an IBC tote as a chicken brooder for day old chickens. Yes, we even kept them in our living room, too.

You can see a picture of the setup on page 107.

Chicken Roosting

Older chickens look to the world above when they are ready to stay goodnight. They instinctively want to roost. There are many ways to ensure that chickens have appropriate roosting places and space. Chickens are woodland natives, and their feet are adapted to sit on tree branches. If you have tree branches of suitable size, strength, and length, these make perfect free roosts. They can be lashed or otherwise secured together or laid across the existing support structure of a house or coop.

Two-by-fours turned on edge also work extremely well. For shorter spans, some will use 1×2s, either purchased or made by ripping 2×4s into two pieces. Personally, I prefer using 2×4 or similarly sized wood, since it holds up better over longer spans. If you have a large number of chickens, making longer roosting runs is easier and better anyway, like our simple H-style roosts.

Simple H Roosts

At some point, we had about 200 to 400 chickens on the farm. Their night-time quarters were the back section of our barn. It takes a lot of roosting space for a few hundred chickens to all have a place to put themselves up at night. The recommendation is about six to eight linear inches per bird, so 300 layers need 150–250 linear feet of roosting space. Also, that many birds produce a lot of chicken poop—just under 50 pounds a year per bird. Every hundred chickens produces two and a half tons of high-powered fertilizer. When you add in the bedding, you are looking at ten or more tons of material each year that needs to be moved out of the chicken house.

For the cost of a few screws, these roosts made from salvaged lumber have lasted through almost a decade of heavy use.

We normally did it twice per year, late winter/early spring and then again in the late fall. (In the summer, manure builds up way less in the chicken house since the birds spend so much more time outside.) At first, we had heavy roosts that couldn't be moved easily. Trying to scoop over, around, and under them made cleaning difficult.

So, I came up with this simple scrap-wood double-H design. We even did triple-H cross bars for a few but noticed the birds rarely used the bottom rung. So we went back to only doubles. These are sturdy enough for a large number of birds to rest upon, but light enough for a single person to easily move out of the way when duty calls.

They are easy to collect and stack in a small area so that you can work freely and unhampered when removing old bedding and manure or doing other chicken-house chores. These were made from free scrap 2×4 so, other than eight screws and a few minutes, have almost no cost. Most run about 8 to 10 feet wide for the cross pieces and use 6- to 8-foot pieces for the verticals. The bottom rung should be around 3 to 4 feet off the ground, and the next at 5 to 7 feet. You need around a foot of space between the top cross piece and the upright's end to ensure space between the wall and the cross piece for the birds to roost.

Angling the tiered roosts is very important—ours normally sat about three to four feet from the wall against which they leaned. Birds should not be above each other so that their droppings land on the ones below. Our deep bedding is what held the roosts in place. In a pinch, you can use a few extra screws, some wire, or some additional scrap wood to secure them to a wall or other spot. I prefer the bedding to keep them in place, since anything extra adds more work when coop-cleaning day comes.

Free Bedding Options

Animals do not live by shelter alone. Animals also need bedding—lots and lots and lots of bedding. Especially if you want to capture all that fertility they create and improve instead of damage your homestead's ecosystems. Many stores sell bedding materials, but they aren't cheap. A small bale of compressed wood chips or similar material costs around five or more dollars. It also doesn't last very long or go very far. Chickens need around six inches of high-carbon bedding material in their coop or enclosure if not free ranged—more is better. Sometimes you can get some or all your bedding material for free—coffee chaff, wood chips, sawdust, and similar materials are often options to replace store-bought stuff. In some cities, they have free pickup places for wood chips, leaves, and similar material. Other times, certain businesses are more than willing to let you pick up loads at their place.

For a few years, a friend of mine did custom woodwork and would produce three or so very large trash bags a week of wood shavings. I traded a bottle of kombucha for the bags each week, which provided all the fresh bedding our hundred or so chickens needed. Combined with the coffee chaff produced by the coffee shop that we picked up grounds from, our animals always enjoyed a nice clean coop, which meant we consistently had almost perfectly clean eggs.

Another time, for a few years a tree company on the south side of town asked me to swing by and pick up excess wood chips from their lot. They would load, I just had to let them know the time a few hours in advance. Had I had a dump trailer this would have been an even better bargain, since it was a bit out of my way. It is just a reminder that if you look and then plan your route through town well, you can turn another expense—gas, wear and tear on your vehicle, and time—into an opportunity to save.

Bulk Buying Animal Bedding

Bulk buying isn't just for toilet paper and other staples and supplies. Bulk buying bedding and other farm or homestead inputs can make a big difference in cost. The local sawmill down the road from my house sells a skid steer scoop of sawdust for $20. It takes two scoops to fill my pickup truck and about an hour round trip. So I am out around 50 bucks and two or so hours of time between loading and unloading for a few weeks' worth of bedding. For $150, another sawmill will deliver a tri-axle load right to my door, which is about 20 to 30 times what I can haul in my truck. They will dump the load wherever I point, and all it takes me is about 20 minutes tops to meet the truck, point out the spot to dump the sawdust, and pay the bill.

Tree companies can be a homesteader's best friend, dropping loads of animal bedding and mulch right where you need it.

For $600, they would bring me a tractor-trailer load, which clocks in at about eight to twelve times the size of the tri-axle load. I would love to get one of those a year, but alas, a tractor trailer can't get up our road, so I make do with the tri-axle loads for the time being. This is an important consideration when you purchase a property—while privacy is great, if your goal is to eventually produce animals or other stuff at scale, the lack of full-size truck access can become problematic.

Just look at the savings both in time and treasure as you purchase animal bedding in bulk. For about three times the cost, I get 15 times the amount in a quarter of the time it takes me to pick up a load with my truck. The only thing is I have to do is plan ahead (it takes a few weeks to get the tri-axle loads delivered) and have cash in hand (remember, cash is king in the homesteading economy).

This also applies to things like animal supplements and so many other homestead inputs: Bulk is almost always better. Sometimes, it makes sense to round up others to split a shipment. We do this for oyster shells, kelp, and some other things that we don't go through fast enough to justify sitting on an entire pallet's worth.

Low-Cost Laying Boxes

Buying laying boxes is a big expense. Metal ones can run well over $200. Wooden and other ones can take a long time to build and are also pretty pricey to buy.

Back when we had a few hundred layers, we needed a lot of laying boxes. Early on we discovered that lining with cardboard made keeping the boxes clean and cleaning them much easier.

This portable laying box repurposed six or so different materials—sheet metal, scrap wood, kitty litter containers, milk crates, and some scrap chicken wire—to make a light, movable set of laying boxes for our current, smaller laying flock.

Milk crates and cat litter containers are low cost or free, often just discarded and lying around in dumpsters, landfills, and other places. They make excellent laying boxes and can be hung in offset runs or on edge in a movable laying box setup. Depending on the number of chickens, we have varied our laying box setup, but milk crates have always served as the backbone of our approach. They are easy to clean, swap out, or remove as needed.

Cardboard Cleanout

We put cardboard in the bottom of our laying boxes to make clean out easier. On occasion, a laying box gets dirty. An egg ends up breaking in the box. Or a chicken poops in the wrong place. Or mucky weather sets in and the chickens track the muck and mud into the coop and surrounding quarters. This is bad for your eggs and a real health risk. Dirty eggs are no fun to handle or clean, so keeping our coop tidy is a high priority.

But keeping laying boxes clean is generally an unenjoyable task—dirty laying boxes often have broken egg debris, chicken poop, dust, and other sorts of detritus all mixed together. The cardboard turns this unpleasant situation into an easy, quick fix, with no gloves required if you are careful (but still wash your hands afterwards). We line our laying boxes with appropriately sized pieces of cardboard and then place the bedding material on top.

When a laying box is dirty, we just grab the upturned edges of the cardboard and remove the dirty set of laying box bedding, cardboard and all. We then toss the whole mess right onto the floor of the coop, where it will compost down into the deep bedding where it belongs. Another option is, if a worm bin is close by, send it straight to a wormy recycling. Replace with new cardboard and drop fresh material into the laying box. This keeps the laying box itself clean while making cleanup easy. You can also substitute newspaper for cardboard—just make sure to use six or so sheets so that when you pull them up they don't tear, dumping the mess in the bottom of the box and defeating the purpose of the paper.

Keeping the Bugs at Bay

One of the less fun sides to homesteading is that bugs call nature home too. Some of these bugs bite. Sometimes, they bite our plants or animals, causing problems. Sometimes, they bite us, causing even bigger problems. Keeping the bugs at bay is an important part of happy homesteading. There are all sorts of things you can buy, but there are a lot of things you can make yourself to help deal with the problem of pests. Over time, it is best to build a homestead full of natural predators and plants that will control pest populations. A few of the below projects can help you do just that.

The Well-Placed Flower and Herb Garden

Some people love those tiki torches and other candles that you burn in order to repel bugs. Instead of buying a bunch of candles or torches though, why not build well-placed pest-repellent herb and flower gardens around your homestead? Instead of having the scent of citronella and other bug-repelling flowers for a few hours, you can have it all the

This small garden space provides many benefits— a lovely view while doing dishes, a haven for beneficial pollinators and predator insect species, a medicinal and herbal growing space, and a strong bug deterrent in an area where we spend a great deal of time as a family.

time at a fraction of the cost. Even better, these plants also help get rid of unwanted pests by attracting pollinators and predators to your place, and many of these plants are beautiful and edible as well. Also, many are perennials. So you plant once and they will come back year after year. You can also propagate many of them easily, providing a source of potential revenue.

Great herbs and flowers to plant include

- Mint
- Catnip
- Rosemary
- Basil
- Lavender
- Marigolds
- Lemongrass
- Chrysanthemums
- Petunias
- Nasturtiums

Attracting Beneficial Birds and Bats

Many homesteads may lack appropriate habitat for birds and bats. This is simple to remedy. A single colony of bats can consume half a million insects an evening. A dozen or so birds can make a significant dent as well. Numerous plans for bat and bird houses abound online. Here is the design we use to make simple birdhouses to hang around our place to reduce pests.

Basic Cedar Bird Box

This is a great project to do with your kids, as it covers a lot of basic woodworking skills but needs very few tools. A mitre saw, power drill, and hammer are the base tool set. Nothing else is required.

The bird houses are beautiful and affordable—we buy number two or similar cedar lumber from our local saw mill, so each house costs us around $3 or so. They can be left unfinished depending on the wood species, such as locust, cedar, and other rot-resistant species, or you can finish them with tung oil or a similar natural finish if you want them not to weather as quickly.

The entrance hole size is very important, as it determines the types of birds that can take up residence in the box. In our area, the ideal size is one and a half inches for the species we are trying to attract. Both the height of the box and the size of the hole are determined by what bird you hope takes up home, so check before you drill.

This is another fun project that can easily result in a small side business selling bird houses to either country or city folk.

One of our goals for 2018 or 2019 is to add bat houses to the mix and see if we can get these prolific pest eaters to call our place home. There are numerous websites that offer great, free bat house designs.

Fly Traps and Tricks

Flies. A four-letter word, especially to cows and numerous other animals. We used to have terrible fly troubles around our place. Over time, we have reduced their annoyance substantially without spending a lot of money or time.

One low-cost way to get rid of flies, especially around areas large animals tend to loaf is by making plastic jug traps.

There are numerous free plans for bird houses available online, so choose one that best fits your skills and the species of birds you hope to attract to your homestead.

Material list
- ▶ Gallon or similar size plastic jugs (milk is great, as is vinegar)
- ▶ Extra pieces of rope
- ▶ Gorilla tape

Steps
1. Cut the top of the jug off about an inch to two below the shoulder.
2. Invert the top and place it back on the jug.
3. Join the seam together with tape, or if you have a heat gun or similar tool, even better to remeld the plastic. But tape will do.
4. Fill the jug with a few inches of water. Add some banana pieces or peel and vinegar. These pieces need to be completely covered.
5. Hang the jugs using rope, baling twine, or the like in trees or otherwise nearby where your cows or other animals loaf. Make sure, if you have goats or pigs, that they can't reach them. Cows don't tend to bother these contraptions, but more curious animals will destroy them quite quickly.

 John's Burgeoning Bird Row

I have slowly been removing trees along our one wood line to allow more sun in the late fall through late winter to reach the south side of our property. For a few of the trees I left the bottom eight to ten feet of the trunk. Across these I mounted a crossbar and have started to adorn it with rows of bird houses. The location is excellent both for watching the birds and for using their beneficial presence to reduce pests in people-heavy areas.

Over time, we continue to build and add additional bird houses on the rail supported by the tree trunks.

Winter-Time Light and Water

In the winter, keeping water unfrozen is difficult, especially without electricity. But, especially for chickens and pigs, 24/7 water and food is completely unnecessary (and creates fire and electric risks in your coops and barns). A chicken needs only a few ounces of water per day in the winter. We take warm water out to them twice a day (in the morning at feeding and then later during egg collection). This way, we don't have to spend nearly so much time fighting frozen waterers or running electricity with all its risks to keep water just above

We hung these in our cow's loafing area so that, while they rested, the flies that so often afflict cattle found a nice place to be reduced to compost.

 ## Ray's Barn Martins Make a Mosquito-Free Homestead

Sometimes, you build something on your place and other passers-by see opportunity where you never did. When farmer Ray built his barn for goats and cattle, with a loft for hay and straw, he never realized who else would call it home.

Over time, passing barn martins noticed that Ray's new barn had dozens of perfect places to build nests. Pretty soon, Ray found himself with 20 or more Martin families scattered across the barn and zero mosquitoes and a number of other pests within half a mile, covering not only his house and some other outbuildings but also the small pond close to the house and barn. He let the visitors become permanent residents and has enjoyed significant bug relief ever since. Be

willing to share with others when you can—you never know when they will turn an unused nook or cranny into a blessing in return.

Amazing engineers, martins can build nests on almost any surface if given the opportunity.

freezing. It also has the added benefit of improving their laying rate and reducing their feed consumption.

Another option, especially if you have a wood stove, is to cook them mixed cracked grains in a pot on the stove. This gives chickens warm food and warm water, combining the two tasks and making the animals exceptionally happy during the cold, short days of winter. Houses that heat with wood tend to have really low humidity during the dead of winter, so this also helps keep sufficient moisture in the air for your comfort and health. Just make sure the cooked grains and liquid cools

 ## Chris Hollen's Barrel of Flies

If you have a lot of fly pressure around your homestead and animals, especially for cows, goats, sheep, and similar larger animals, Chris Hollen came up with an excellent fly killer.

Material List
- 30- to 55-gallon barrel
- Catchmaster Pro Giant Fly Trap (one roll will do an entire barrel)

Apply the fly traps around the barrel. Place the barrel *outside of* but close to areas that contain larger animals (cows, pigs, sheep, and the like). Placing it close to watering stations is especially effective. The bugs that bother your animals will quickly land all over the barrel. Once they get going, all their friends think it is a good place to land and you are in business.

You can let chickens and ducks pick the tape clean for easy protein, as friends who have used this say the fowl generally don't get themselves stuck on the sides when grabbing a snack. Some do report a bird will occasionally become stuck on the barrel, but it is a pretty rare bird.

Credit: Chris Hollen

The barrel fly trap takes fly control to a new level of portability and simplicity.

My Three-for-One Fly Food System

One of our homesteading principles is that we want to get as much value as possible from every single task we have to do and want nature to do as much work for us as possible. For example, we plant comfrey because it not only attracts pollinators, serves as a useful medicinal herb, builds soil, and is a good animal feed, but it also suppresses weeds and looks beautiful. One action, many benefits.

With flies, you can turn this pest into a source of protein for your chickens or similar animals while also breaking the insects' reproductive cycle and radically reducing their pressure. All it takes is getting them to reproduce where their larvae will never see the light of day.

Years ago, we discovered that flies had an affinity for coffee grounds when a few buckets got left sitting out for about a week before we dumped them into the compost field. Each bucket contained hundreds upon hundreds of fly larvae. The chickens thought this was fantastic! These are thousands of potential flies that will never get their wings or live to reproduce.

Now, we make coffee-grounds or manure-based compost piles in frames near our chickens. We get the coffee grounds in town from a number of coffee shops that are more than happy to see someone collect them instead of tossing them into the trash. We already have to go to town a few times a week anyway, so it makes our trip more productive economically and ecologically, because coffee grounds have a tremendous amount of fertilizer value. But we don't just use them as a fertilizer. We use them as a fly control and a worm food as well. Many points of value from a single action.

To make a pile, we layer coffee grounds and some rotted wood chips together in a frame.

After a pile is built, we make sure it stays sufficiently watered. After a week or so, we check the pile's status. It should have lots of small maggots and other critters crawling throughout it. We give it another four to ten days, then pull the frame to let the chickens scratch through the pile. Our chickens get free food. Our farm gets great compost. And the various pests get removed from the reproductive cycle. The material can then be finished as compost or tossed into our worm bins, providing another round of free chicken feed and producing an improved compost that not only is better for plants and soils but is also easily sellable at an excellent profit margin.

And it all starts with a few free bags of coffee grounds!

some before offering them to the animals, as you don't want them to get burned or injured. If they are used to getting just cold, dry rations, this is a real risk, and, just like with a person, burn injuries, especially internal, are not to be taken lightly.

Instead of using electricity for their water, we save it for a blue spectrum (daylight) LED light on a timer in their coop. This helps to keep their lay rate up during the dark days of winter. The greatest energy/environmental cost of a chicken is in the feed. To feed all winter long while getting few to no eggs just doesn't make environmental or economic sense. An LED light on a flock of ten to twenty birds is an immense environmental savings. You can even run such a light off a small solar panel. Chickens need about 12 to 14 hours of light per day to lay, so adjust your the timer accordingly every few weeks based on how much daylight you have and how many additional hours they will need to still produce eggs.

We don't worry too much about giving the chickens a "break," as they already get that every year during molting for three or so weeks and, unlike confinement birds, are not forced into constant, daily laying.

There is a lot more that could be said about caring for animals, but animals alone don't make a farm. Our journey is only partly done, so let's move on to another part of the homestead, shall we?

Climbing Rope

Even a surprisingly simple item like a used length of climbing rope has many uses for a homesteader. A friend of mine who runs a tree company gave me a few lengths of used rope, each around 40 feet or more in length. Reputable businesses replace their ropes fairly often for safety, but they are still quite durable and useful around the homestead. We have been using one section for almost six years. Following are a few ways rope has helped us.

Try reaching out to a local tree company, not just for chips, but also to see if you could have some of the discarded lengths of rope for reuse at your place. You will not be sorry to have them on hand.

Tire Swing

No homestead is complete without a tire swing. All you need is rope, a tire, and an appropriate tree. It will provide hours of fun that you may even find yourself sharing with your kids on occasion. The climber's rope also makes great additions to obstacle courses and all sorts of other such creations for you and your family.

Bale Toppling

Many a truck can fit two round bales of straw or hay in the bed. The trick becomes how to get them off, since they will end up tightly packed together. A simple solution is to use a length of rope shimmied between the bales, with one or two people on each side, to pull it off.

Without a tractor or skid steer, how do you move tightly packed round bales off a truck or trailer? We found climber's rope lets the kids do the job.

Snow Cleaning

Our area suffers from incredibly inconsistent snow and winter weather. We have had years with less than a few inches and days that dropped multiple feet. Keeping rarely used

and expensive snow removal equipment on hand makes little sense. I love this solution Curtis Stone popularized a few years back for cleaning snow from various structures, especially high tunnels, but also from barns, and other smooth-roofed outbuildings.

You need a section of heavy-duty rope—our old climber's rope does an excellent job—and a length of easier-to-throw, lighter rope. Tie a loop about a foot in length into the heavy-duty rope every four feet or so. Tie the lighter rope to one end of the heavier. If this still isn't enough to allow you to toss the lighter rope over the structure, use an old "foxtail" toy (shown below) to help you toss the test rope over.

During winter, we just leave the rope draped over one end of our high tunnel so that, if we get bad weather during the night,

An old section of rope becomes a useful way to remove snow from structures

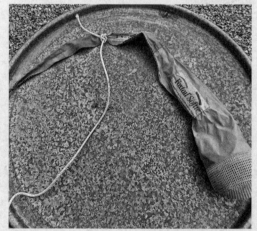

Even kids' toys (like this Foxtail) have more than one use when solving homestead problems.

we are not trying to get the rope into place in the midst of it. Realize that you may end up having to remove snow from a structure at three a.m. The last thing you want to try doing in the dark is getting the rope in place.

To remove snow, have one person holding the rope on each side of the structure and gently seesaw the rope back and forth. The knots will knock the snow off quite efficiently once you get the hang of it. This works great as long as the structure has an intact roof with no vent pipes or other protrusions. We have used it on our house to clean excess snow the few times it was called for, but it was a great deal slower and required a lot of care to not damage anything sticking out of the roof. It was still easier and safer than trying to get up there, though.

Load Securing

Our rain water catchment at the barn was a craigslist find. At the time, I had no trailer large enough to accommodate this 1,500-gallon tote, but my truck bed did just fine. The only problem? No ratcheting tie-down or other such contraption long enough to secure it was in my possession. All were severely short, reaching at best a third of the way around the tank. Thankfully, I had an entire length of rope in the truck as well, which worked perfectly to secure the tote in place for the 20-some mile drive back to our place.

I hope the above shows you how something as simple as a section of used rope can save you multiple hundreds of dollars if you can imagine all the different ways you may use it both for work and play.

[3]

In the Barn and Workshop

One man's trash is another man's treasure.
— HECTOR URQUHART's introduction to 1860's
Popular Tales of the West Highlands

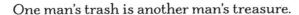

ONE DAY, many, many years ago, an acquaintance of mine at school came bustling by, clearly excited about something. Theology? Possible girlfriend? These were generally the two major concerns of my classmates. But his was neither. His mind was on a dumpster. Well, actually, it was *in* the dumpster, but in a good way.

Just around the corner from the school a house was undergoing renovations and an addition. The builders were tossing any unused/unneeded/unwanted materials into a roll-off dumpster that sat right along the street. This was back just a bit before the housing crash of the early 2000s, so it makes sense why they cared so little for the materials. As he was driving through the neighborhood, my friend noticed the ends of some 2×10s and 2×12s sticking out. He pulled over, peeked over the side, and was amazed to see not some measly leftover cut-offs, but full-length boards by the dozens.

An hour later, we had my friend's pickup truck bed filled with over $800 of perfectly good lumber. Mainly 2×10s and 2×12s, most ranging in length from 8 to 14 feet but lots of other great materials as well. These boards sell for over $10 each at the store, and here was a dumpster overflowing with them for free. We could have gone back for at least one, maybe two more loads. It was a lesson that stuck with me long after. Often times, perfectly fine stuff is thrown away for no reason at all. You just have to know where to look for it.

Almost two decades later, we are still using this wood for projects around our place.

Low-Cost Shelves, Quick Shelves

Some of that lumber is still in use today, about two decades later. I recently had a lean-to put on the west side of our house to provide some much-needed additional storage. The uprights were a perfect place to put in some shelves. But how? Strong ties! For the cost of the strong ties, some bolts and some screws, combined with the free wood from over a decade ago, we now have lots of shelving right alongside the lean-to on the house.

Notice that the shelves are vertically offset. This allowed us to use the shorter pieces of free wood we had on hand, rather than trying to cover a long span and having to buy expensive lumber to do so.

Tool Storage

Tool storage doesn't have to be fancy or expensive. A few spare barrels can hold a lot of long-handled tools. A few buckets can hold smaller hand tools. Some long, heavy-duty nails and free space allow you to make hanging racks for smaller ones. For odds and ends, you can use cat litter containers or other buckets hung off the nails or screws. Barns and other spaces often have unused spots that, for just a few bucks, make great places to store tools and other stuff.

"Oiled Sand" to Stop Rust and Protect Your Tools

Rust. Few things are as damaging to your investment in hand tools. An easy way to help protect your investment involves using buckets or barrels filled with "oiled sand."

To make oiled sand, you will need a five-gallon bucket, a bag of sand (or about four gallons' worth volume wise), and a half-gallon of the oil. Put the sand into the bucket. Pour the oil evenly across the top. Allow it to sink in and then away you go.

Some places suggest using motor oil or mineral oil for the oiled sand. I just can't fathom putting that kind of stuff into my soil. Instead, I use a vegetable-based bar chain oil—the same stuff you use for a chainsaw—or a food-grade mineral oil. Some of my friends also use canola or similar vegetable oils. Given their low cost and availability, stick with one of these safer, more organic options.

Keep these buckets close to where you store your tools. When you come in from work, first clean and, if needed, rinse off the tools, then plunge them into the oiled sand a few times. Shake off the sand, wipe down the tools with a spare rag also kept

A few cross bars and some nails can turn unused space into easy tool storage. A few cat litter or other buckets can store odds and ends.

close by, put the tools away, and go about the rest of your day knowing they are well protected from rust. This is one of those homestead tricks I wish I had been told about sooner.

Storing All Your Stuff

There is an endless array of stuff to store on a farm or homestead, from animal feed to tools to toilet paper and out-of-season clothes. (We have a large family, so we keep a lot of toilet paper around in case of emergencies.) How can you store it without buying or building expensive storage space and equipment? Let's explore some options.

Barrels help protect stuff stored in barns and other outbuildings while keeping things tidy and easy to move around with a dolly or similar equipment.

Barreling Along

At one time, we had no local source for non-GMO animal feed. So I had to get in three or more tons of it at a time. A ton of animal feed is 40 50-pound bags. One hundred and twenty or more bags of feed takes a lot of barrels to store at six bags a barrel. I watched for and found a commercial kitchen for a restaurant chain that produced a lot of used, food-grade barrels, and I scored 20 plus of them for around seven dollars a steel barrel. They had about three or four a week to get rid of, so every other week I would swing by and get another set. Why every other week? My truck could hold about eight of them, so this minimized my time, fuel, and mileage investment, while maximizing my return for the trip. I purchased a total of about 25 to 30 barrels. This gave us plenty of space to store feed, while also leaving a number of extras for other stuff.

Most of these barrels are still with us today, save ones I sold off at two or more times what I originally paid. We have never regretted having extra, low-cost ones around for storage or projects like the compost tumbler. If you find an especially good deal on homestead stuff, don't be afraid to pick up a few extra if you have the cash and space to do so. If you don't end up using them, you can often sell them off at a tidy profit.

Storing Animal Feed, Supplements, and Soil Amendments

Proper storage of animal feed and supplements, along with soil amendments and many other items, is very important. These tend to be some of the largest annual expenses for homesteaders and farmers. Moisture, mice, and so much else will cause all that money to go to waste. Proper storage is how you help stop the loss.

Depending on your scale, barrels are generally the first and best storage option for smaller operations—a few dozens chickens, perhaps a few hogs. If the feed and supplements are kept bagged, a standard 50-gallon barrel will hold about 250 pounds (five bags) of feed. Unbagged, a barrel will hold 300–350 pounds of animal feed. This is a great amount for the average homestead, as it is best to use animal feed as quickly as possible because of nutritional loss. Certain nutrients in animal feeds degrade quickly, especially during the summer, so we try to have no more than two weeks of feed on hand at a time. We keep bagged minerals and supplements in a separate barrel or two.

HOMESTEAD HOLLER — Old Chest Freezers for Animal Feed and Storage and So Much Else

Years ago, before we had our own place, some friends from church needed a farm sitter for a few days. We were eager to help, thinking it would be a great introduction to homesteading. They had goats. If you know anything about goats, you know where this is going. We had no idea what we were getting ourselves into.

During those adventurous first few days as farm sitters—which included a number of run-ins with the goats—we noticed a number of old chest freezers around their place. A few were for animal feed. Another stored miscellaneous stuff—clothing and personal items.

Chest freezers are great for many years of reuse like this, especially if they are not exposed to weather. They are secure, pest proof, and relatively easy to clean. They also provide some temperature stability for their contents, which is a big bonus for storing animal feed—if they are not in direct sunlight, they won't heat up much during the day. While our land won't permit an attempt at this idea, we have seen people bury their old chest freezers to use as root cellars as well. If you go this route, I would suggest first removing the compressor, wiring, and anything else of recyclable value that isn't needed once it is in the ground.

At the end of its useful life, a chest freezer should find a final home at an appliance or metal recycling center. Don't do what a some country folk do and toss your used stuff in sink holes, hollers, and other such places. Nature doesn't want or need our used stuff.

With 20–30 chickens and two or three hogs, a 50-gallon barrel of feed will last us about two weeks. During winter, when we want to have extra on hand in case we get iced or snowed in, and because the animals go through a bit more per day because of the cold, we set up a second barrel and keep an additional five or six bags on hand. This ensures that we don't risk running out of feed, especially when the animals would most need it. During winter, nutritional loss is lower and slower as well.

We like to keep our barrels, especially the steel ones, out of direct contact with the dirt. This helps stop condensation inside the barrels and also helps protect them from rust. We stack them on pallets in our barn to extend their life and stop moisture and mouse issues underneath the barrels. Especially if your barn is at risk of occasional flooding, stacking stuff up on pallets is low-cost insurance. Our barn has flooded a few times over the years, and we have never regretted not having to worry about light flooding since everything of importance is four or so inches above the ground because of the pallets.

However you store animal feed also has to keep it protected from other hungry animals. Raccoons, possums, and many other opportunistic critters will try to get into your feed barrels if left unsecured. If a barrel lacks a locking mechanism, a simple solution is to toss a heavy rock or concrete block on top of the lid to keep the feed in and the hungry fiends out.

Hay and Straw

Hay and straw—the stalwart friends of farmers and homesteaders. If you have animals of any size, you will probably need hay and straw. Many people use them as a mulch for their gardens and growing spaces or as bedding in their animal places.

Given how the average person treats these precious resources, an onlooker wouldn't agree with that assessment. In my area, thousands of bales of hay are stacked on the ground, side by side, and left outdoors all fall and winter long until needed. Sometimes, they sit there year after year, rotting down to nothing for no good reason, a waste of fuel and fertility.

How much loss are we looking at? When exposed to weather or moisture, hay loses over a third of its feed value. Bales stored outside for more than six months are basically wasted time and money, along with reduced-quality animal feed. They may also begin to produce molds, which can possibly sicken your animals. In many areas, round bales cost around $30-$50. Improperly stored, that results in 10 to 20 or so dollars of loss per bale. Just look at the dollar bills going goodbye when you store your hay and straw outside.

Bringing them under cover isn't enough, though, even though it is a big improvement. Ground contact can cause the bottom of bales, or the bottom few rows for square bales, to degrade, while also creating an ideal insulated habitat for mice and other pests to multiply. Like firewood, straw and hay need to be kept up off the ground and in a covered location.

The good news is this isn't all that hard to do, even though a lot of folks don't bother to do it. There are a few easy ways to get it done.

First, pallets are perfect for keeping both round and square bales off the ground. If you are going to stack bales many rows high, look for the strongest pallets you can scavenge. A square bale weighs anywhere from 40 to 75 pounds. An average pallet is made to bear 1,500 to 2,000 pounds. Oak pallets can often hold al-

most twice that, around 3,000 pounds. If you want to stack square bales extra high, stick with the strongest pallets you can source or use 4×4 rails instead.

If you find you can't get pallets, you can make rails to roll or stack the bales on out of 4×4 or similar lumber. This is another time that seconds and other reclaimed materials with defects find a new life around the homestead. Such lumber may not be safe for building or framing with, but it is great for storage and similar projects.

If you are really pinched and don't even have any suitable wood you can use, another option is to find some appropriate-size tree branches to use as rails instead. Our woods are riddled with decent-quality and -diameter branches or small saplings that would work fine for such a purpose. We have seen many a homesteading friend substitute tree branches and delimbed saplings to store straw and hay on.

Other Stuff to Store in Barrels

Animal feed isn't all we get in bulk. We have barrels full of toilet paper. Barrels full of wool. Barrels full of bags of extra or out-of-season clothes. If you have a big family, extra storage space in the form of barrels is hard to beat for properly storing all sorts of stuff that doesn't need temperature stability. If it can take temperatures down to freezing and up to a hundred, it can go in a barrel! For clothes and similar items, you may need to pack them into bags first, along with adding bay leaves or similar herbs to protect them from moths.

The other caution is that you also need to keep the barrels out of direct sunlight and weather. A dark-colored barrel, steel or plastic, can heat up to 120 degrees or more in direct summer sunlight. This is not only hard on the contents but a possible fire hazard. So if you use barrels for storage, make sure their spot is shady and weather protected, like your barn.

Simple Barn Loft or Shelves

If you need to store a large amount of lighter stuff in the barn for a time but don't want to go through the trouble and expense of building a full barn loft or shelves, this is a simple solution we have used many times over the years to make additional space. Sometimes it stayed up for only a few weeks. One has been sitting this way for about two years. Either way, it works very well to give a bit of extra, lofted storage space without taking a bunch of time or material to do it.

This project is best done with 4×4 or similar lumber for any span that is 8–12 feet in size. Most barns are built on eight- or ten-foot centers for the uprights. Take your measurements, and then use the appropriate lumber.

Material List

- ▸ One or two sheets of plywood or similar material
- ▸ Three 4×4s, determined by the distance you need to span
- ▸ Two 2×4s or 2×6s x side distance

The amount of weight the loft can hold depends on a few factors. First, the distance of the span. Second, the size and species of wood you use to support the plywood and cross the span.

The larger the pieces of wood, the more the loft can support. The longer the span, the less the loft can support. Softwoods, like pine and cedar, are able to support less weight than hardwoods like oak. We tend to limit our lofts to 300 or so pounds of bulky-but-hard-to-store homestead stuff that doesn't fit well in barrels, bins, or other types of storage. Extra cases of mason jars. Bar-chain oil and two-stage gas-mix back stock. Currently unneeded chicken feeders and similar equipment.

We use this loft mainly as work space and light-duty storage. Since it uses no screws or nails, it is easy to both assemble and disassemble. The 4×4 or similar wood is often extra wood we keep on hand for more

permanent projects, so it costs us nothing to temporarily put it to use and, when no longer needed, place it back in the stored lumber stacks. The cross pieces that the span pieces sit on were already present in our barn, since it was originally a set of horse stalls.

When placing the running rails, be sure to do it on the *inside* of the area where the loft will be. This allows the span pieces to sit much farther over the runners, making them far more stable and unlikely to slip off, taking your loft with them. The span pieces should sit at least two or three inches over the edge of the rails for this reason.

Let There Be Light

Electricity. Oh, so convenient. Until it causes a fire. While in homes this is rare, this is a real risk with barns and outbuildings, especially since many of these buildings are not wired properly. Barns and outbuildings are not just home to all sorts of flammable substances and farm inputs like hay and straw. They are also home to rodents, who find electrical wiring an irresistible chew toy at almost any time of year. My friend Joel Salatin mentioned to me years ago that one of the reasons they pulled all the electricity out of their barn-type buildings was because of the countless times he heard about neighbors' barns burning down that had electricity in them. As long as hay was dried right, you rarely heard about it otherwise.

When we moved to our place, the first thing I did was remove the electricity from both the barn and another outbuilding. Someone had hastily cobbled together what seemed like little more than leftovers from a number of home appliances and underrated extension cords to put electricity into these places. While they didn't fully succeed or fail in their efforts, it did make both buildings incredibly unsafe. Removing it was absolutely necessary and is a good homestead lesson. Sometimes, you have to *undo* before you can *redo*. It also created a new problem—the barn was now, even in midday, terribly dark.

One reason people want electricity in their barns, with endless wires that mice and other problematic pests can nibble on, is for light. It is hard to work with animals in the dark, dungeon-like environment

Hay is for Horses—and Horrible Fires

(It's been a while since we were talking about hay.) I want to take a moment to remind you of the importance of properly drying and storing hay. If you are a long-time farmer or home-steader, please hear me out—just last year a farmer friend with over 20 years' experience watched $20,000 go up in flames from improp-erly dried and stored hay.

When hay is cut, the grasses, legumes, and other plant materials contain a great deal of moisture. It needs to dry out before being baled. Sometimes, weather goes wrong and hay gets rained on. Or sometimes we get in a hurry and bale it too soon. Once baled, all isn't lost. The hay just needs more time to dry. But sometimes, the bales get rained on as well, rehydrating the hay.

If the hay has enough moisture, the microbes in the bales go to work, digesting the plant material and producing a tremendous amount of heat, just like in a compost pile. As the reaction gets going, the moisture may get used up so quickly that the bales catch fire, and once one goes, all the adjoining ones will go with it. There are a few ways to avoid this sad story that afflicts thousands of farms and homesteads each year. First, you can purchase a hay moisture meter to test your hay and ensure that it is sufficiently dry. Properly dried bales will be around 18–20 percent moisture. Make sure to test them in multiple spots and test multiple bales, as bales are not uniform in density or composition, nor are fields.

It is also good to test your bales with a probe/compost thermometer. These have a much longer shaft to allow you to find the temperature of thick or deep materials. The two- to six-week window after baling is the danger zone for when hay is most likely to go haywire. Testing some bales a few times a week should result in internal readings of around 125 degrees. If you notice bales that are higher, keep a close eye on them—twice daily monitoring or more. If bales hit 150°, it is time to get them *out* of your barn or storage building and into a place where they can prop-erly dry before they burst into flames.

Second, you can add preservatives to your hay. The easiest way involves sprinkling salt on top of the bales. Salt is a natural drying agent and preservative. There are other options on the market—acids and other agents—that you can use if you ended up with no choice but to bale hay that wasn't sufficiently dry.

Last, make sure you store your hay properly. As we discussed on page 95, it should be in a cool, dry place, with good air flow if possible, and not in contact with the ground, nor exposed to weather or excess moisture. Follow these simple rules and a hay fire will never be in your future.

that many barns provide, especially during late fall and winter when daylight is short and weak.

We found that a few simple changes can turn a barn into a beautiful workspace with no need for electricity with all its costs and risks. Also, if you need tools up in the barn, go cordless, or run an extension cord from a nearby properly rated outlet. If your place has one properly located pole with power, a few high-quality extension cords are a fraction of the price of properly installing electricity in a barn.

Low-Cost (Sky) Light

The lowest-cost way to dramatically improve a barn's friendliness is through replacing a metal ridge cap with a clear poly skylight ridge. This one change alone allowed us, even in the dead of winter, to work in the barn at most hours of the day with no need for supplemental light. Unlike doors, which sit on only one side of a barn and thus capture far less light for far less of the day, a skylight ridge cap gives light all day long, and even at night during strong moonlight. Unlike trying to put in

Make your barn nice and bright with a simple skylight ridge cap replacement.

skylights, which are fairly expensive and, unless done right, can lead to roof leaks and other problems, this is a simple DIY replacement that, depending on your barn, will run you under a hundred bucks. All you need is the proper width skylight ridge cap, proper screws for attaching it to your roof, and then the proper sealant, if any, to keep heavy winds from pushing rain or weather under the gap and into your barn, along with a cordless drill and the right bits to remove and replace the old ridge.

Hands-Free Light

One thing we have bought for every member of our family is a headlamp for farm work. Flashlights have limited usefulness to a farmer or homesteader. They take up one of your hands, and generally you need at least two, and preferably three or four for a lot of tasks, especially at night. If you are like us, you only have two, and generally when trying to capture an escaped chicken or finagle some other farm task at night you need them both. If you get some headlamps, make sure you get ones that have both white and red light settings. The red light has two benefits. First, it doesn't ruin your night vision. Second, it doesn't confuse your body or disrupt its rhythms. This allows you to get to sleep (or go back to sleep) more easily when your work is all said and done during the wee hours of homestead life.

Everyone in our family has their own headlamp that we keep powered with rechargeable batteries.

Small Changes to Structures that Let in Lots of Light

While the first low-cost change to make to a barn or similar building is substituting skylight ridge cap for metal, in some large buildings or certain barn designs this may not be enough. If you still need more light, a second upgrade for metal-roofed barns could involve replacing some of the metal with skylight panels, especially on the south-facing side of the roof to let additional light in during winter.

Adam Barr's Better Brooder

Along the south side of an outbuilding, farmer Adam Barr replaced some of the metal siding with polycarbonate panels, turning his once dungeon-like brooder area into a delightful place for day-old chicks to grow up. This has had a number of benefits. First, it gives the animals access to lots of natural light, which is better for their health and growth. Second, it creates free warmth, reducing his heating costs for his brooder area. Third, it makes a far more inviting area for him and his crew to work in. Small chicks are often difficult to check on, so the additional light helps make inspections and spotting issues a lot easier. Such is the power of light to make us and our animals happier and healthier and our lives easier.

This simple replacement of metal siding with poly panels made working with the chickens much easier and their brooding quarters far more animal friendly.

 **Catch Rabbit Poop Right into the Worm Composter**

*One creature's waste
is another creature's food.*

Nature is amazing. What we often see as waste is actually one animal or creature taking a resource that is unusable to almost all the others, taking a portion for itself, and then excreting the excess to bless others in the system. This is especially true of animal manures. Animals eat all sorts of stuff, make use of and absorb only a portion of it, and then give back the rest, improved and enriched, in the form of manure and urine.

Rabbits take large amounts of green matter, through some arcane alchemy turn it into meat, and also return it to me in the form of many pounds of rabbit poop that is rich in all sorts of nutrients, including nitrogen. This poop, when mixed with bedding, makes a perfect worm food. But who wants to pitchfork rabbit poop into worm bins—homes designed for containing and raising composting worms—when you can just let the rabbits put it there for you?

Instead of a Raken (rabbit and chicken house made popular by Daniel Salatin), we have a Rabin, a rabbit and worm bin house. By building our rabbit hutch tables to the right width, we can place worm bins right below the hutches. This means no pitchforking is required, just the occasional adding of carbon and bedding and turning the worm bins like we already have to do. The rabbits tend to also ensure the bins have sufficient moisture as well. The Rabin area stays clean this way. My kids get one less chore. The farm produces great compost for use or sale.

Worms eat our rabbit poop and give us compost in return, instead of making an extra bi-weekly cleanup chore for Abby.

Pallet Barns and Outbuildings by the Buchlers and Jameses

What if you don't have a barn yet and don't have the money to build a large, permanent structure? Or maybe you are heeding wise homesteading advice and waiting a few years until you better know your land before you begin plopping down buildings hither and thither?

You can do what many of my friends have done and build temporary barns and outbuildings using pallets and other materials. Both Rachel James and Randy Buchler have done a number of pallet structures, and thousands more examples abound beyond these.

The one consideration with a pallet structure is ground contact. Wood in direct contact with the soil will rot and, depending on conditions, can rot out fairly quickly. One way to address this is to build a pallet structure with a base of treated lumber or on concrete blocks to provide some level of moisture protection. Another is to keep the structure to a workable size. A pallet structure is usually best kept under 16×16 feet in size, depending on your roofing options and skills. A narrow, longer structure, something like a 12×16 is far easier to build and roof and is what I would recommend for those starting out. It would provide a tremendous amount of storage space if built and set up right,

but the roof span is such that a simple shed-style roof would alleviate a lot of issues and dangers with snow load and other such things. Depending on its use, you could go with a dirt, gravel, or even a pallet floor.

Credit: Rachel James

Credit: Rachel James

Credit: Randy Buchler

You can build an almost endless array of outbuildings, animal houses, and other useful structures using pallets.

The Many Uses of IBC Totes

Another highly versatile and low-cost reuse item that every homesteader should at least take a gander at is the intermediate bulk container (IBC). These generally run between 250 and 350 gallons in size and consist of three parts—the plastic tote itself and then the cage and pallet that make the tote and its contents easy to move with a tractor, forklift, or skid steer. The pallet may be wooden, plastic, or metal. The latter two are far more durable. If wooden, remember the warnings given previously about wooden pallets, and check to see if it is heat treated or chemically treated.

In Chapter Five we will talk about using them for rain catchment, but that is only one of an almost endless array of uses for these totes. So let's look at some other ways these can be repurposed to improve your homestead.

Storing Kindling and Small-Sized Wood Scraps

The pallet cut-offs and slab wood we use as a mainstay to help heat our home have an added benefit—tons of smaller pieces that are perfect for kindling. For years, though, we found all these small, oddly shaped pieces difficult to store. They certainly don't stack well and filling a pallet-type structure with them was always a mess. While barrels worked fine, the amount we picked up from our neighbor's sawmill filled far too many barrels for my liking. One day I realized that an IBC tote with the top cut off was the perfect place to store a large amount of these pieces, and the problem was finally solved.

IBC totes are an easy way to store kindling and other small, non-stackable materials.

Baker's Green Acres' Animal Feed Storage

Many years ago I found myself at Mark Baker's farm up in Michigan. It was there that I first began to realize that IBC totes have many uses beyond mere rain catchment.

Especially if you have a larger operation and go through animal feed quickly, these totes are great for storing and moving around bulk (as opposed to bagged) feed. A tractor with forks or similar equipment makes using these for feed storage far easier, but I have seen growers using them even if they lacked these options and just scoop out feed as needed. The top can be completely cut off, or you can cut off about 80 percent, leaving just the back edge of plastic. This gives you a "hinged" lid to help protect the feed from certain critters who might be tempted by its uncovered presence.

If you need to secure the lid, you can drill holes in both the lid and the rest of the tote, using wire to tie them together and keep larger critters from climbing up and in.

I have seen a few farmers go even further and add actual hinges and a latch to the cover for this reason. If you leave the top uncovered, the tote will need to be stored some place where the contents are safe from both weather and unwanted guests. Many farms that use these for feed have guard animals or lockable buildings, so make sure if you go this route you can make it work.

Worm Compost

For years, this has been my favorite, most common use for these totes. We produce about four to six tons of worm compost per year. The totes make worm composting easy, since they provide an almost ideal home for worms. The large size makes management and care of our worms a lot easier—the totes are more temperature, moisture, and habitat stable than smaller worm compost-

Our first use for IBCs was to provide space to scale up our worm composting operation.

ing systems. Since they come with a drain, as long as the tote is sited properly, with the back edge a few inches above the front, causing the tote to angle forward slightly, they drain easily.

Animal Shelter or Chicken Brooder

Another nifty use we have seen for these totes is as an animal shelter. Since the totes are so easy to clean, such a use is very appealing. While we have not personally used them for cats or dogs, many a chicken (meaning, a hundred or so at a time) has spent a week or two in our IBC tote brooder. Quite a few times the brooder found itself in our living room, and we went to sleep to the sound of hundreds of chirping chickens each night. The young children especially delighted each morning to wake up to the endless chirping that accompanied breakfast.

By removing some of the metal cage and cutting a door into the tote, they can serve as

An old IBC tote cut in half became the perfect place to start out batches of chicks.

houses for larger animals, such as dogs, cats, and even small livestock.

So Many Other Uses

I have seen many other uses for these totes. For small-scale pasture fertilization and application, it is easy to rig up and attach a multi-head sprayer system to the tote's drain valve. Toss the tote on the bed of a heavy-duty pickup truck, fill with your fertilizer mix, drive to the pasture, orchard, or other location, set the spray volume to desired application rate, and drive. Some people with heavy-duty four-wheelers have a trailer they set them on and then spray from. Whatever you have can work if it can bear the weight.

Also, these are popular for aquaponics, sometimes even in basements or garages in urban environments. Just realize that if something goes wrong, 250–350 gallons of water in a basement or garage can make a real mess. An acquaintance had the float valve on his malfunction while away on vacation. His setup was in his basement, and the water ran from the unfinished to finished section. Let's just mention that homeowner's insurance may not cover such fiascos. It didn't for my friend, so be mindful of not just what you use these for but where as well.

In areas where mail and package delivery won't go all the way to the homes on more rural roads, I have seen people make one of these into a package depot where deliveries can be dropped, and then they can collect orders when they pass by.

Hopefully, this section gives you a glimpse into how versatile and useful IBC totes are beyond catching and storing rainwater. At the same time, IBC totes and many other containers have certain risks that we need to talk about for a few minutes before you go gung ho over them.

IBC Tote, Plastic Barrel, and Other Material Safety Concerns

A lot of previously used items are perfectly safe for homestead reuse, but many are not. You need to learn how to tell what is and isn't safe, or if and how you can make something safe again for you, your soil, and your animals.

First and foremost, you need to know what a container previously contained. Was it filled with a natural dish soap or synthetic

If a container comes with its label intact, this can help you determine its safe uses around your homestead.

herbicides? Motor oil or olive oil? Unfortunately, a lot of items don't come with easy-to-read labels. Instead, they come with chemical names usually, but not always, listed on a label on the side of the tote, barrel, or container. These chemicals will each have an MSDS—a material safety data sheet. You can find the MSDS for any chemical by searching online for MSDS + name of chemical. You have to spell it exactly as it appears on the label. There are well over 60,000 chemicals in use in the modern economy, between industry, consumer products, agriculture, cosmetics, and so much else.

Many very different chemicals have very similar names and spellings, so make sure you write each one down and type it in correctly. Sometimes, the difference of just a few letters may be the difference between life or death.

Also make sure not to confuse *trade* names with *chemical* names. Trade names are what you see in stores on product labels. Things like Tide or Roundup or Penzoil. Chemical names are 2-Aminoethanol, glyphosate, synthetic hydrocarbons, and polyalphaolefins. Trade name products are usually made up of many individual chemicals, and the MSDS for each may need to be pulled to make an informed decision.

Once you know exactly what you are dealing with, you are finally in a place to decide if and how to use the container and what remediation efforts might be needed first. Some chemicals are never safe for contact with food or water. Some can be neutralized and removed or washed out. Others are of little to no concern at all and some dish soap and a hose will get your tote or barrel into service with ease.

How to Avoid the Hassle Outlined Above

Sometimes you may not need to look into the MSDS. If the containers are rated certain ways and were used only in compliance with the rating and the contents are edible or food safe, then this is the fast track to knowing it is safe to use for any purpose on your farm or homestead. One of these ratings is "potable." Potable means safe for drinking water. That is, you can fill it with water and drink the water and not die or become deathly ill or contract some terrible disease or get covered in discomforting boils or some other such thing. Many totes, barrels, and other storage containers are either not rated as potable, either because of what they contained or because of the materials they were made from. Many more, because of what they previously contained, are no longer suitable for use as potable water storage. Some contained industrial chemicals or cleaners. Others have contained pesticides and petroleum-based products and byproducts.

The second rating to look for is "food safe" or "food contact approved." Like containers rated potable, as long as only similarly rated stuff was inside them, they are generally good to go after a thorough but simple cleaning with dish soap and water.

Some chemicals can be cleaned out of totes even though the chemical is neither potable nor food safe. I once picked up a shipment of totes that contained a concentrated form of hydrogen peroxide used to clean printing press lines. Another time I took a shipment of totes that contained an enzyme used as a feed additive for livestock. The totes needed a thorough cleaning to ensure safe use for water, food, or similar purposes, but both the MSDS and a bit of searching confirmed the chemicals wouldn't leave a residue or infiltrate the plastic in a way that rendered the totes permanently unusable. All it took was some good-quality dish soap and a hose to get them back into service. Me, my kids, and my animals are all boil and pestilence free.

What if You Can't Make It Water or Food Safe?

If you can't get a tote back to water- or food-safe storage, there are still a lot of uses for it. You can use the cages that come with the IBC totes to store larger pieces of firewood or as a hay feeder, for instance. But it definitely limits your options and their uses. I personally don't keep non-food-safe totes and containers around because I don't want to have to remember which are which or have to keep the kids or other helpers straight on the subject. Given that a small mistake here can have big costs, it isn't worth the risk to save a few bucks. If you do keep them around, make sure you clearly mark and differentiate them from your food-safe ones, both for your own sake and the sake of others.

What About a Cage With No Tote?

What if you come across a cage and pallet with no plastic tote? Or, what if the plastic tote is no longer usable? You are in luck! Either way you have a cage that can be repurposed for many good tasks. Here is a quick simple list of ways to possibly use it, sometimes with, sometimes without, the pallet:

▶ Firewood storage—the cages are easy to fill with firewood, and also easy to move if you have a tractor with forks or skidsteer. Keep the pallet, and make sure it is in good enough condition to take the weight of the wood inside the cage.

▶ Square-hay-bale dispenser for cows, goats, lambs and similar animals—it provides great access while keeping them from getting into the hay.

▶ Kid's play house.

▶ Leaf and other compost—these work really well for making leaf mold.

▶ Person bucket on tractor lift. I

wish I had known about this option when we were finishing our high tunnel, as it is a much safer option than standing on a bare pallet alone. You can make it safer by securing the cage and pallet to the lift using a few ratcheting tie downs or rope.

Depending on your application, make sure you check what the pallet is made from (plastic, metal, or wood), and if it is wood, that it is safe for that purpose. (See page 35 for details.)

Credit: Dave Perozzi, Wrong Direction Farm.

If the tote isn't safe, you can still use the cage for all sorts of creative things, like the Dave did for drying and storing firewood.

[4]

Outfitting Your Home

In the US, roughly 40 percent of the average family's income is gobbled up by their home. While some of that is hard to avoid, there is a good bit of room to realize savings in what is generally the biggest category in your budget. We will talk about some areas, like utilities—heat and water specifically—in Chapter Five. But there are many other things we can do to reduce the cost of our living spaces or the expense of caring for them

Trading Time for Money

A number of years ago I got a call from a friend at a buying club I had helped start a few years before. They were frustrated by the behavior of one of their members. The club recently began offering fresh milk, and this woman was already getting milk from the same dairy farm. Pickup at the club was an extra dollar or so compared to on-farm pickup, and she had flatly refused to get the milk at the club because of the "incredible additional cost," as she put it. Instead, once per week she would load up her five kids, drive 45 minutes each direction, or around 70 miles total, to save $5 at best on three gallons of milk, instead of just grabbing it with all her other stuff at the club.

Many of us don't realize that, like this lady, we are "penny wise and pound foolish." Her milk trip was easily costing her $20 just in gas and mileage. When you factor in the value of her time and trying to load multiple small mayhem-causers into a car, well, it is clearly a bad deal to save a few dollars. To successfully farm or homestead on a shoestring, you need to make good decisions. Sometimes DIY is the way to go, but sometimes used or borrowed or bartered is a much better deal.

HOMESTEAD HOLLER 📢 Paul Aud's Coolbot Refrigerator Conversion

Credit: Paul Aud

A lot of homesteaders start out by purchasing a run-down or dilapidated property, trading sweat equity to achieve their homestead dream. The Auds are one such family. One day, when I was talking with Paul, he mentioned they were doing the kitchen renovation on their home. New refrigerators large enough for his family easily ran over a thousand dollars for about 20 cubic feet of storage. Instead of sinking money into something undersized, Paul decided to build a much larger walk-in refrigerator into their kitchen using a Coolbot for temperature control. For slightly over double the cost of that 20-cubic-feet fridge, the Auds now have 800 cubic feet of storage (well over 30 times more than a standard refrigerator). They have ample space to save and store all sorts of seasonal foods and other things that they would never have had space for otherwise. The Coolbot also costs about the same or only slightly more than a pair of refrigerators to run, another bonus to both their bottom line and the environment in general.

Ten to twenty times the storage of a refrigerator for only twice the cost.

Sometimes, it just makes sense to buy it, sometimes used, sometimes new. But you will only get so far and survive so long if you make bad decisions like the lady above.

Making Your Own Cleaning Supplies

A bottle of basic spray cleaner or glass cleaner in the store costs around three to five dollars. You can make it at home in about two minutes for about 50 cents. Given that homesteading is at times both messy and mess making, you can save a fair bit of money making some of your own cleaners without having to spend a great deal of time. Here is the basic one we use most around our house. The only thing you need beyond what is called for in the recipe are some good-quality spray bottles.

Basic Spray Cleaner

Adapted from *Clean and Green: The Complete Guide to Non-Toxic and Environmentally Safe Housekeeping* by Annie Berthold-Bond.

Combine in a spray bottle and shake well:
- 2 cups hot water (We heat filtered water on the stove top, but you can also use very hot tap water.)
- 2 tablespoons white vinegar
- 1 teaspoon borax
- a squirt of natural dish soap

Basic Glass Cleaner

Adapted from *Clean and Green*, this recipe works so well and is so easy to make. Combine in a spray bottle and shake well:
- 2 cups water (We use filtered water.)
- 3 tablespoons vinegar
- a squirt of natural dish soap

Also, for some cleaning supplies, if you are able order in five-gallon or larger sizes, like we do with dish soap, when the container is empty you end up with a useful bucket instead of another piece of trash or recycling.

What is a Coolbot?

Imaging taking a standard window air conditioning unit and using it to create a refrigerated environment. A Coolbot makes this possible at a fraction of the cost of compressors and other larger cooler technologies. These are popular among small-scale farmers and some businesses, but they also present some unique applications like Paul's for homesteaders as well.

Storing Your Surplus: Sloped Bucket Storage from Scrap Wood

Like many homesteaders and large families, we buy in bulk to bring down food costs and also to provide some provisions in case of emergencies. We generally have as many as 20 five-gallon buckets full of foodstuffs—salt, sugar, various beans and grains, along with other such dry goods.

Five- and sometimes six-gallon buckets are often the go-to for storing bulk goods on the homestead. Bulk buying can save a family thousands of dollars a year on food and many other expenses. A pound of organic raisins at the store? $4 or more a pound. A 30-pound box? $75 at our buying club, less when on sale. A pound of organic oats? Around $2. A 50-pound bag? $43. A three-pound bag of organic apples? Five to eight dollars. A bushel box, just $44. When you are feeding five or more people, the savings add up quickly.

Large dry-goods purchases—salt, sugar, grains, beans, and similar foodstuffs—need proper storage, which usually involves five-gallon buckets. These buckets can quickly take up a lot of space when left on the floor, and they became a chore to try and sort through when looking for a needed ingredient. A simple solution is to create a tiered,

Some free lumber made these large shelves cost just five dollars and a few hours of work, and they allow us to store large amounts of various dry goods.

slightly slanted, bucket storage rack from scrap wood. We built two of these for our buying club, and each would hold 15 five-gallon buckets in about 40 square feet of space while allowing full access to each bucket. That's efficient storage! Gamma seal lids make getting in and out of the buckets doable even for small children, while easily keeping them fully sealed. Also, we use labels on the buckets' lids so that, with just a glance, everyone knows what was inside each one. That way, the lid and bucket can be reused for other items just by removing and replacing the sticker, instead of writing on the bucket or lid itself and creating confusion as to the contents and a mess on the surface.

Material List
- ▸ 2×4s and 2×6s, 2×8s
- ▸ Screws

A few tips for this project. First, I like the buckets to have a 20 or so degree pitch. Much steeper and this setup doesn't save much space or make the buckets easier to access, much shallower and when you open the tops stuff falls out. Items could theoretically be stored underneath the shelves as well, especially if it is longer-term stuff that you don't need to access often.

This is an all-scrap-wood shelf—even the angled supports that hold the cross pieces on which the buckets will sit. The two sides of the shelf are exact copies of each other, just mirrored.

Once the two sides are complete, all it takes is installing the cross pieces. You can use narrower wood, such as 2×6 for the rear cross piece, while the front cross piece will need to be 2×8 since more of the bucket leans along the front facing side.

Can You Store What You Grow?

After a few years, we have got fairly good at growing food. Along the way, though, we found a flaw in our plan. Our place has ample space to grow food, but the infrastructure for storing it is poor. There is no basement. No root cellar. No concrete-floored outbuilding, no temperature-stable outbuildings. The ground has rock just a few inches to a few feet down, so putting in an outdoor root cellar isn't an economically viable option either.

If you find yourself in a tough spot such as this, you have a few options. First, store outdoors whatever you've got that doesn't care about extremes of cold and hot. Extra clothing, kids' toys, craft and other supplies and many other things, if stored in a covered area in sealed totes, don't overly care what the weather does. Potatoes and other produce care a lot. Garlic braids make a great home decoration, and let you hide your garlic in plain sight instead of trying to find a place to squirrel it away. Things taking up space in closets and under beds can go outdoors, making room for bins of winter squash, boxes of potatoes, and containers of other produce. Instead of keeping magazines and movies in a coffee table, a rustic wooden storage bin full of potatoes could be the center piece in your living room.

Early in our marriage, Jessica and I would purchase a few hundred pounds of local winter squash each fall, mainly spaghetti and butternut. The lower closet in our apartment became squash central, with two wire racks storing all this delicious goodness. Sure, it meant we had to own less other stuff and clothes, but that has made life easier for us over the years.

 One for the Books

How do you handle limited storage space? Especially if you don't know how long you will be at your place? Some farmer friends of ours in Indiana decided to go portable and purchased an old semi-trailer at auction. For a fraction of the cost of building something, while making it easy to move when they finally settled on a new farmstead, this bought them a few years' of time and space to deal with the stuff an expanding family creates.

Indoors and out, there are creative ways to store stuff, whether personal belongings, produce, and grains or other items. You just need to discover and take advantage of them.

Turning Trash into Treasure

Jessica has done a number of projects for us over the years that were enjoyable for the family, made our home more inviting, and saved money in the process. This usually involves items whose first purpose in life is no longer workable, and thus they get the "Joseph Had an Overcoat" treatment, resurrected through her skill and inspirations.

Here are just a few.

Rag Quilt: Upcycling Old Flannel and Jeans

With five kids, we go through a lot of clothing, especially jeans (boys and jeans…) and bedding. A lot of good fabric is left on a pair of kneeless jeans or badly stained flannel shirts and sheets, and all this can go into making lovely rag quilts. Unlike a traditional quilt that requires the use of a long-arm quilting machine or many hours of hand sewing, a rag quilt is much easier to make and needs only a regular sewing machine. Jessica made two simple yet beautiful rag quilts for our sons' beds out of old jeans and old flannel sheets, and they cost us almost nothing other than the thread and needles for the project, along with the time and love one put into each.

Draft Blockers

Another nifty use for old jeans is turning them into draft blockers that prevent cold air from coming in under a door. This is done by cutting a long portion of a jean leg, sewing off one end, stuffing it with various materials (we like buckwheat hulls), and then sewing the other end

These help around a few old doors during the coldest parts of winter to keep heat in and wind and cold out.

Mounting this bag on a heavy-duty coat hanger makes it easy to keep it handy during laundry duty.

closed. The pants may need patches in a place or two, which can add to the decorative and aesthetic value of the draft blockers. Depending on how tall your family members are, you may need to make two or more per doorway or other place you plan to use them.

Clothespin Bag

Recently, our boys discovered the need for their clothes to be, occasionally, clean. As a family of seven, we go through a lot of laundry each week. Whenever possible, we hang it outdoors to dry. Since we hang our clothes out, we need and use a lot of clothespins. Hundreds and hundreds of them. How do you handle so many?

An old towel that no longer delivers in the bathroom can become a useful clothespin bag. You simply cut the towel, fold it leaving extra fabric on one side, sew the sides to make a large pocket, and then sew another seam up the center of the large pocket to make two smaller pockets. Then Jessica wisely folded part of the towel over a clothes hanger and sewed it in place so it has a handy, built-in hook, making the bag easy to move, store, or hang on the line when hanging laundry.

Homemade Organic Pillows

You spend roughly a third of your life on a pillow. Eating organic food is great, but good bedding also matters. Organic pillows in the store are very expensive. We just happened to have a friend whose parents had sheep. I asked him what they did with the wool. He said, "Nothing," and that they would love for someone to make use of it. That wool has filled many a pillow and child's toy in our home over the years.

We have another friend who gave us bags of wool from their sheep, too. You may need to trade a little labor for some free wool, but many sheep farmers would love to have help at shearing time, and you can not only get some wool but also learn a valuable skill at the same time. Combine the wool with some buckwheat hulls—not expensive by the 50-pound bulk bag—and some organic fabric and you have yourself a half-dozen or more pillows.

With organic cotton outers and wool and buckwheat hull infill, these pillows are lovely to sleep on.

HOMESTEAD HOLLER — Life Without a Dryer

At some point during our homesteading adventure, our clothes dryer went splunk. When we moved into our new place we only had a washing machine. The house improperly vented the clothes dryer into the crawl space, so instead of spending money on a new dryer and having to totally redo a long dryer pipe run (another improper installation issue and a big fire risk), we decided to save some money and space and stick with a wood stove and an outdoor clothes-drying system instead. As a big family, it requires some planning, but the savings are large. An average load of laundry costs about 50 to 60 cents to dry. In some locations, the cost is as high as a dollar a load. For us, it equates to about $200 or more a year in savings, not including the greatly reduced risk of house fire or having to spend a few hundred dollars to redo our laundry area to get a shorter, safer run for the exhaust.

Also, line drying doesn't just save cents, it saves your clothes. They last far longer when line dried. All that lint that you have to fish out of your dryer every few loads or risk a deadly house fire? That is your clothes losing a few fibers at a time with each tumble. You also save on bleach or similar extras as well, since the sun keeps clothes looking nice and smelling fresh through its natural bleaching action.

Pieces of Scrap Pipe to Move Heavy Objects

A few scrap pieces of pipe make moving heavy things a lot easier.

Ever have an item too heavy to move? Like a large piece of furniture, chest freezer, or other stuff like that? A simple way to make moving it easier is to use pieces of pipe to roll the object along. This approach goes way back—we are talking six thousand or more years—and was used all over the world to move massive pieces of stone to build all sorts of walls, castles, monuments, and even the pyramids. The Egyptian pyramids weren't built by aliens but homesteaders.

We keep a pile of scrap PVC pipe around for this purpose, so that whether outdoors or in the house, if we need to move something heavy to a new place, we can do so with even just small kids to help. Generally, you want a minimum of five pieces of pipe, so that the object is always balanced on three or four and you have an extra pair or more to place in front to keep your momentum.

Cinder-Block Shelves

If you have some spare concrete or cinder blocks and access to some extra wood, these are easy-to-make, versatile shelves for storage. During my days in seminary, a number of my friends had bookshelves

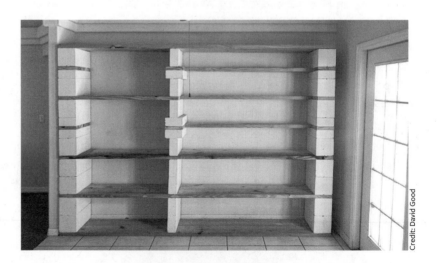

A cinder-block shelf can embody both function and beauty if well built.

Credit: David Good

made in this style. If you have access to rough-cut lumber, it gives them an appropriately rustic look. Since no screws or nails are used, you can easily repurpose the wood to other projects later, along with the blocks.

Two Pairs of Shoes

Years ago, whenever I came across a pair of shoes I really liked, I always purchased two pairs, sometimes even three. This past year, I purchased four pairs of the same shoe. Partly, it was because I always look for shoes on sale and when a good deal comes up, I stock up. Partly, I hate changing shoe styles only slightly less than I hate changing clothes. But there is another benefit. Instead of wearing a single pair every day, I swap shoes. After a long day of wear and use, shoes compress under all the weight and stress. Overnight is often not long enough for the materials to rebound. By swapping shoes every other day, they last far longer and provide better comfort and support. So if you can swing it, when a pair of shoes you like is on sale, snag two pairs instead of one, and swap them out every day or so. Your shoes will treat you better and last a lot longer.

Paint Your Entire House for Under a Hundred Bucks?

Because the house on our farm doesn't have much inherent value (it is an old double-wide mobile), it is hard to justify spending large sums of money on it. So what do you do to cut costs while not sacrificing quality? When we moved in, the entire house needed painting. A 2,200-square-foot house has an immense amount of wall space to paint, probably close to a thousand dollars' worth or more. I have a friend who is a professional painter. He pointed me to a possible solution to such a significant expense—mistinted and returned paints.

Our local city has a few dozen paint shops, including paint departments at big box stores like Lowe's and Home Depot. So, when we were in town, I would drop in and ask to see any mistints and returned paints they had on hand. These normally sell new for $30-$50 a gallon, but as mistints or returns they are available for around five dollars a gallon, sometimes less. Some are for outdoors. Some are for indoors. Some are latex. Some are acrylic. The colors, type, and quantities vary week to week and even day to day. So we realized it was going to take a few months to get all the paint we needed.

We would collect colors we liked well enough, and for larger rooms, collect three or four gallons that would mix well (based on color and type of paint) to provide enough paint to make one new color in sufficient quantity to do the entire room. Once we had enough, I would mix small, equal portions of the various cans to make a test batch to ensure that the final mix worked and provided an acceptable color. Sometimes, stores have five-gallon-bucket mistints or returns, which are also great if you need to paint a very large space in a single color and don't want the trouble of trying to mix your own.

Mistints are a great way to get paint for all sorts of projects, especially if you are not overly picky about the color. Personally, I would rather have an extra thousand dollars for many other things in life. I also don't care much what color the rooms and other projects around the homestead are as long as Jessica likes them well enough. It took a few months to get what we wanted, but the savings made it more than worth it. Also, the shops were right near where we were going in town anyway, so we weren't adding a bunch of extra time and miles either. We traded an hour or so for multiple hundreds in savings. Best of all, the paint looks great!

Cutting the Cost on Good Clothes

The best time of year to buy things is when no one else is. If you can get used to shopping out of season, you can save significant amounts of money. In the late summer, all sorts of spring and summer farm and garden supplies go on clearance. A friend once found bags of peat moss and coir (coconut fiber) for $1.00 bag in early fall, instead of the usual $12.99! Another friend picked us up a half-dozen seed-starting trays for 25 cents each on mid-summer clearance.

This is also true for the type of heavy-duty clothing homesteading requires. My winter farm jacket normally runs $140. Instead, I purchased it in late winter on clearance for $45. We are keeping our eye on a swimming pool right now, at the beginning of winter, waiting for online retailers to clearance them to rock-bottom prices to make room for next year's inventory.

Learning to buy out of season can save you a fortune. Socks, hats, boots, jackets, and so much else often goes on sale for 50 to 80 percent off after Christmas. *Also, don't be afraid to bargain.* I have asked to see store managers and dickered down additional discounts many times on close-out and out-of-season items.

Sometimes, it was on ripped or damaged bags of compost, potting mix, or other growing supplies at a big box store. They were the leftovers of the mark-

Our animals have yet to complain about these imperfect cattle panels, while we continue to appreciate the significant savings.

downs and had many holes, which is why they were passed over repeatedly by other customers. An offer of a buck a bag let me claim them all, compared to the clearance price of three to four dollars. Other times, bargaining netted me a store coupon or some other savings. When I needed some cattle panels, I asked if the store had any seconds that were dinged or bent here or there and thus couldn't be sold as new. Some stores pile such materials in a separate spot so they can either recycle or otherwise dispose of them. The manager knocked 30 percent off the price of each panel, and I am pretty sure neither my cows, kids, nor hogs can tell a difference. Remember, the worst thing that can happen is they say no thanks to your offer.

Finding Appliances and Furniture

Furniture represents an immense expense for a large home or family. An average room takes three to four thousand dollars retail to fully outfit. That was never in our family's budget, but we also didn't want to buy low-quality particle board junk that would fall apart and need replacing every few years. So how do you affordably outfit a home without compromising on quality? This story about my Amish friends shows one way.

When the Amish Come a-Calling

For many years, I have had Amish and Menno-nites friends across our home state of Kentucky. Some supplied food to our buying club. Others were locals whom I got to know and become friends with by purchasing things from their businesses or having overlapping interests. A few have already popped up in the book, but not Elvin. He was a cheesemaker down in southern Kentucky who supplied our buying club for many years. He also badly beat me at chess a few times during visits down to his farm.

One day I received a letter from Elvin, but it had nothing to do with cheese. Elvin had a problem. His oldest daughter was due to be mar-ried the following summer. Thus, their slightly overrun farmstead needed a thorough cleaning so they could host a few hundred members of their family and community. "Overrun" was an understatement. On top of having a saw-

mill, dairy, and cheesemaking operation, years before a friend of Elvin's had asked him to store and sell some of his solid oak Amish furniture. Hundreds and hundreds of pieces were piled all over Elvin's place. In the cheese parlor. Under the wraparound porch. His entire well over two-thousand-square-foot basement was piled to the ceiling with it. Tens of thousands of dollars worth of furniture tied up almost every free inch of space in his house, workshop, cheese shop, and other outbuildings.

Then the furniture maker suddenly died. After the furniture maker's death, Elvin had

We hope that this furniture will one day find a home in our kids' homes.

bought everything at a hefty discount to help the family. After two or so years, he had sold a fair number of pieces, but he was still overrun by over a hundred some pieces. Pieces that needed to go pronto to make space for the impending wedding. He wanted it all gone and he was willing to give an unbelievable price—about $30 a piece for furniture that normally retailed from the mid-hundreds to low thousands.

Elvin asked me if I wanted to buy it all. At the time, my wife and I lived with 2.5 kids in a two-bedroom apartment totaling maybe eight hundred square feet. It just wasn't an option. We did have many friends who would also love to upgrade their furniture. So we hatched a plan.

Find five other families and do the great furniture buy. Sixty-some thousand dollars worth of furniture for six thousand dollars, split among the six families for a thousand dollars each.

That was a memorable day. The shipping truck pulled up into the parking lot of a friend's business. (We gave him a few free pieces to let us use his warehouse for a few days to do the deal.) A few hours of unloading were followed by an NFL-style furniture draft until all the pieces were spoken for. Well, almost all. It was so much furniture that after families had laid claim to dozens of pieces, some finally said, "Enough! I can't take any more!" It was that good a deal!

About half the furniture in our home is from that day. It took some sanding and finishing. (We all also went in on a bulk order of tung oil and a few other items so we could finish the furniture.) But at the end of the day, we ended up with over $10,000 worth of solid oak furniture for pennies on the dollar. Almost every family had extra pieces. Some sold those at a tidy profit per piece over what they paid. By the time we sold or bartered our extra pieces when we ran out of space or motivation to finish them, we ended up breaking even or even being a bit ahead.

One of the best parts is knowing that many of these pieces will, Lord willing, end up in the homes of our children and grandchildren, a testimony to their quality and craftsmanship but also to the value of relationships and taking advantage of sudden opportunities, even if they require some creativity to make happen.

Other Great Places to Grab Furniture— If the Amish Don't Come a-Calling

Early in our marriage, our family outgrew our dining room table fairly quickly. Especially because I ran my own tutoring business, that table often had to seat our family plus a steady stream of high school, middle school, and occasional college students and their math, science, and other textbooks. My wife also needed a place to store our nicer dishes and serving bowls and platters. One day on craigslist, I saw a family selling an oak dining room table set and a china hutch/buffet. Brand new, each retailed for well over two thousand dollars. The family was selling both on craigslist for $600. They took a cash offer of $400.

As I write this, we need to upgrade a bathroom with a new vanity. New, with no sink or faucet, it now cost almost $200 for a basic model at most stores. On craigslist, I have found vanities for around $100 or so, ready to go with top, sink, and, sometimes, even the faucet included. At the Habitat for Humanity ReStore, you often find them for 50 to 100 bucks.

With outfitting a home, a few things are important. First, realize that all the credit cards in the world won't do you any good for getting the best deals. American Express may be everywhere you want to be, but it's not with the Amish. Visa isn't going to get you very far on craigslist, at least currently. Cash is king. If you don't have it, you can't buy it (and probably shouldn't if you don't have the cash for it, anyway).

Second, patience and the ability to pounce are irreplaceable. We waited many, many months for that table and buffet and paid for them a few hours after they popped up for sale. Our bathroom doesn't need repairing ASAP, so we are watching for a good deal on the materials with similar patience. Since we are not in a rush, we don't have to take whatever offer comes along or pay retail. Since we have cash in hand, we don't ever have to miss a great deal. Since we have extra space to store stuff, we can slowly collect materials until fully ready for a project. Position yourself for success early on in marriage, life, and as a homesteader, so that over time things get easier, financially and otherwise.

Greenhouse Plastic

Another item we have found many additional uses for beyond the original one is greenhouse plastic. This heavy-grade, UV-stabilized plastic does a lot more than extend your growing season. High tunnels now dot the entire landscape of the United States, each generally covered with this style of plastic. On occasion, perfectly good plastic has to be replaced because of weather and storm damage. The storm damage often leaves a few small tears here and there or along one side or edge, while the bulk of the plastic remains serviceable. Or an animal, such as an owl, cat, turkey, or some other critter, will rip up an area of the cover, necessitating its replacement while leaving a large piece available for other purposes. There are a lot of ways to get extra life out of these pieces before they get a final recycling. Also, if you put up a high tunnel or similar structure, you will often end up with some extra pieces of plastic once the process is done. These smaller pieces can serve a lot of useful purposes before heading to the recycling center.

Covering Wood Piles

I dislike tarps a great deal and have been rather unimpressed by vinyl billboard tarps as well, even though a lot of homesteaders speak highly of them. After six to twelve months, both begin to degrade and fray into innumerable, annoying pieces that, if not dealt with quickly, spread and sully large areas.

On the other hand, greenhouse plastic hasn't ceased to impress me, including as an excellent way to cover wood piles and other outdoor stuff that needs protection from weather. With wood especially, the plastic helps keep the pile warmer,

Help keep your wood dry by covering it with greenhouse plastic.

which aids drying. Unlike tarps and vinyl billboards, if it gets a hole, it doesn't begin to fray and fly off in a million frustrating pieces. Instead, some heavy-duty duct tape saves the day quite easily and allows the plastic to continue its service for many more months. Also, unlike covering wood piles with sheet metal, the plastic is easier to secure and less dangerous if it gets free and airborne. We have found that standard greenhouse, six-mil plastic (plastic thickness is given in "mils," short for millimeters of thickness—the higher the number, the tougher and more durable the plastic) or higher is very durable for this and many other tasks.

Solarizing Weeds

If there is one technique I wish I had learned many years sooner for my growing spaces, it would be this one. Solarization harnesses the sun to reduce weed and weed seeds by trapping incoming solar energy underneath the cover and concentrating it in the top few inches of the soil. At the same time, it helps knock back soil-borne pathogens and pests. It works best when done within 30 to 45 days of the summer solstice, but this depends on your location. The farther south you go, the greater the window where this technique will work.

When done properly, solarization renders an area clean from weeds while helping feed and build the soil at the same time.

When done properly, solarization removes 95 percent or more of both weeds and weed seeds in the top three to six inches of the soil. For us, this has reduced weeding labor by about 80 percent in areas to which we applied it the past year. If you practice low- to no-till growing, it makes a significant dent in weeds for many years to come and does so in just a week or two. We kept a piece of scrap plastic specifically sized for strip solarizing single beds so that, as early and late spring crops give way to summer, we can quickly clean up beds before getting the next crop going and growing.

The solarization process is simple. Clean an area that needs to be solarized of debris and stubble that may puncture the plastic.

We run a lawn mower over the area, then handpick out any remaining sticks or other stubble. Water, if the soil is dry, until damp and moist but not waterlogged. Spread the plastic over the area, then bury the edges in a trench about six inches deep on all sides. Wait five to fifteen days, depending on time of year, weather, and other factors. It will need more time if it is cool, cloudy, or rainy, less time if it is hot, sunny, and clear weather. Remove plastic and enjoy a far less weedy growing space.

Low Tunnels

Don't have the money or space for a high tunnel? Well, why not make a miniaturized version to extend your growing season? Low

In more temperate climates, many crops thrive with just a little added protection, such as carrots, greens, and other cold-hardy crops. A low tunnel provides this protection at a low cost.

tunnels can be made using EMT conduit or PVC pipe. A grower can easily switch from greenhouse plastic cover in cooler seasons to shade cloth/floating row cover in the warmer seasons. This not only can extend your growing seasons and increase your harvest substantially but also protect your plants from particular pests as well, especially brassicas like cabbage and kale and other non-flowering crops that don't need pollinators.

If you go with metal EMT conduit, you will need a simple tool known as a hoop bender to make the arches. You need to decide exactly what size you want your growing beds, as the benders are width specific. We went with the six-foot bender, as we prefer wider

beds that allow for companion planting and sprawling plants to have sufficient space without overrunning adjoining beds.

Here in Kentucky, I and many friends have successfully overwintered many cold-hardy crops using low tunnels, allowing us to store carrots, kale, and other crops in the garden for continual harvest all winter long. It also allows late fall plantings to overwinter in place and get a head start come spring so that you sometimes have carrots ready by Easter.

Cold Frames and Hot Boxes

Instead of using glass, you can use small, leftover pieces of greenhouse plastic to create cold frames and plant propagation boxes. A bit of scrap wood, a set of old hinges, and some nails or screws, combined with just 15 or so square feet of leftover plastic is enough to make a large cold frame or hot box to give plants longer life or an earlier start on the growing season. If you are in a really cold climate, you can even double-layer the plastic on both the inside and outside of the box to increase the insulative value.

[5]

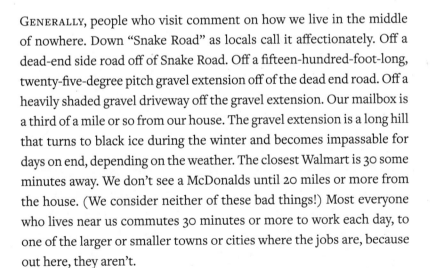

Becoming
More Self-Sufficient

The greatest fine art of the future
will be the making of a comfortable living
from a small piece of land.
— ABRAHAM LINCOLN

❧ ——— ❧

GENERALLY, people who visit comment on how we live in the middle
of nowhere. Down "Snake Road" as locals call it affectionately. Off a
dead-end side road off of Snake Road. Off a fifteen-hundred-foot-long,
twenty-five-degree pitch gravel extension off of the dead end road. Off a
heavily shaded gravel driveway off the gravel extension. Our mailbox is
a third of a mile or so from our house. The gravel extension is a long hill
that turns to black ice during the winter and becomes impassable for
days on end, depending on the weather. The closest Walmart is 30 some
minutes away. We don't see a McDonalds until 20 miles or more from
the house. (We consider neither of these bad things!) Most everyone
who lives near us commutes 30 minutes or more to work each day, to
one of the larger or smaller towns or cities where the jobs are, because
out here, they aren't.

Yeah, we are in the "middle of nowhere." If power goes out, nobody's
coming to check on us or put it back together until most everyone else
is good to go again. If the hot water heater breaks, it is a hundred or
more dollars and a few days to a week before we can get a basic service

call. If things go bad, it is on us and our neighbors—along with the skills and stuff we have on hand—to stay well and safe and see us through to the other side of things. Thankfully, we have many good neighbors just a few minutes' walk away. This is a pretty big part of homesteading—the ability to prepare and provide for yourself and those in your community come what may.

And trust me, come it will. We have had temperatures so cold they froze our water *and septic* systems solid (minus 30° for three straight nights in central Kentucky!). We have had snow so deep that helicopter was the only way to get off our property (24 inches in eight hours—so much snow that it even closed the Interstate in our area for over a full day and stranded countless motorists). We have had windstorms so bad that they took out the *entire region's* power grid (over a million or more homes and businesses, with some people not getting their power restored for over a month). We live in a region that isn't prone to earthquakes, hurricanes, fires, and so many other things that you may have to deal with on top of the possibilities mentioned above. The world is not always a welcoming place.

How will you prepare for all these possible problems you may face as a homesteader? What can you do about them, other than take precautions and make preparations? Or neglect to do so and become a part of the problem when things like the above take place?

You may not be that far out, but everyone can benefit from being more self-sufficient and resilient, even if you are more urban or suburban than rural as a homesteader. Big cities lose food, power, and water quite often, sometimes even more than more-isolated country folk. The past decade of hurricanes and other calamities has shown just how quickly and surprisingly such events can unfold, and just how much suffering and disruption they can unleash on those unwilling to take steps to prepare.

Self-sufficiency also helps protect you against other types of problems that might pop up. If you are in debt, have to buy all your food, and have no other income options, a lost job is a big deal. But what if you have a large garden, food put up in the pantry, animals out grazing in the grass, plants around the farmstead where you can forage, and a low-cost or debt-free homestead? Well, that takes a bit of the bite off the bad news. It buys you some space and time to make good decisions moving forward, instead of having to panic to make ends meet and keep food on the table. Indeed, for us, this degree of self-sufficiency has time and time again allowed us to weather the storms that a family of seven surviving on one income faces fairly regularly.

So in this section let's look at low-hanging fruit in the field of self sufficiency and resilience. We will look at food, water, heat, and other basic needs on the homestead, things that can get you through a bad few days or even a bad few months if you do them right.

Food

An old saying goes, you can survive three minutes without air, three days without water, and three weeks without food. While that may be true of food, it isn't really good advice. Yes, you may survive. But you won't be happy doing it! Especially if you have kids—they are even less happy with insufficient provisions than adults, and a lot of adults I know don't do well when calorie restricted even voluntarily. I can't imagine what they would be like when forced to forgo food, coffee, and other goodies under duress. I think I would rather be trapped with a couple of wild boars.

So let's talk about how you put food on the table affordably and put those foods up—plant, animal, and otherwise.

Deciding What to Plant

How do you know what to plant in your garden? This question came up a few weeks ago in conversation with a good friend, himself a very successful homesteader with many years experience. He was asking how much space did I think it would take to grow all the vegetables for a single person for a year. I pointed out that I thought it was the wrong question. Instead, I said a person should grow based on three rules:

1. What they will eat.
2. What is most valuable to grow given their space and time constraints (either dollar or calorie wise).
3. What does well in their space (based on soil, pests, and the like).

For instance, while potatoes may be easy to grow, organic potatoes are inexpensive to buy, even locally, compared to fresh greens. Also, fresh greens lose much of their nutritional value during transit. Fresh greens from the garden are a bigger cost savings and bigger nutritional benefit. So, I would prioritize what would save our family the most money and provide the most nutrition first.

At the same time, though, if you are tight on money or have a big family, potatoes represent an immense amount of *calories* compared to greens. It would take ten pounds of greens to give you the same amount of calories found in a pound or so of potatoes. The potatoes are also easier to store and save for later, something the greens have going against them. You won't be snacking on spring lettuce in the dark days of winter, but you may be making stews with your root crop stores.

You could do sweet potatoes, which not only give a great yield calorie and space wise but also give you greens. The young, small leaves of sweet potatoes are an excellent green even during the hottest parts of the growing season when most others bolt or bow out under the long, hot days of summer.

Why Do We Grow What We Grow?

You may find yourself asking, "Well, what *do* you grow then?" We grow a lot of things! We grow potatoes and sweet potatoes. Why? They grow easily for us and produce reliable, bumper crops of a high-calorie filler food for the entire family. We also grow lettuces because these are expensive and our family of

Sweet potatoes are a staple crop at our homestead, producing both green and root to feed our large family, along with extras for our animals to enjoy.

seven can eat 30 dollars worth in a week if given the opportunity. Some expensive crops in the store—like fennel and leeks—grow easily and don't take up much space, so we grow these to add variety to our diet and reduce our food costs.

Onions are inexpensive from other good farmers in our area, so while we grow some, we don't worry too much about this high-labor crop anymore. We have significant squash vine borer issues in our area, so we try to grow summer and winter squashes but enjoy only limited success because it's hard to control this pest organically. Peppers and tomatoes take up space every year, and lots of peppers end up diced and frozen for winter use. A few specialty crops that now dot our growing season, such as ginger and turmeric, have great nutritional and medicinal value and easily fetch a fair market price for the extras. Green beans are another staple that produces reliably and often ends up with extras to sell or store. But given their labor requirements and the low cost of organic frozen green beans, we don't worry about putting up too much extra.

We also have a lot of perennials—blackberries and raspberries, strawberries and serviceberries. Overall, we grow what works best for us, given our soil, preferences, and time investment. A big factor with everything we grow is that it is *efficient* to grow in terms of time, calories, and cost. We also grow only things we know our family will eat—no eggplant and few hot peppers here!

Every year, we grow a few new vegetables, both to expand our family's palate and our options for what to produce. Fennel was one of my favorite 2017 plantings, along with ginger and turmeric.

The quintessential homestead animal. Low cost, easy to keep, and hard to keep alive!

My daughters' second favorite homestead animal. Highly productive but almost completely dependent on store-bought feed.

You will need to experiment and figure out what you and your family like to eat, figure out what grows well given your soil and location, and figure out what makes the most sense financially and food security wise.

What is the Best Animal?

Oh boy, I know. I am opening a can of worms, and not the kind you eat. The best animal on a homestead is a hard call. Honestly, there is no such thing as a "best" animal. Every animal has pluses and minuses, benefits and drawbacks.

Chickens are relatively easy to keep but something everything under the sun likes to eat, and they generally require supplemental purchased feed. Some will breed easily, many will not, requiring annual flock replacements. They are also easy to butcher.

Rabbits? Well, they breed like rabbits. But most require special store-bought feed because they are so different from their wild cousins that they can no longer survive on pasture and forage alone. They are also pretty high on the predator "What's for dinner?" pecking order. Like chickens, they are another easy animal to turn from dancing around to dinner plate.

Pigs are fantastic at turning discarded and expired stuff into delicious meat—and also require an approach to fencing that makes places like Fort Knox jealous. Cows turn grasses and other green material into grass-fed beef, a true miracle. But they take a lot of time to do it, and require fencing only slightly below what pigs need. Also, cows require hay and other supplements to stay healthy, along with shelter of substantial size, especially over winter.

Cows present far more risk as well. A dead chicken is an annoyance. A sick or dead cow is a sizeable economic loss and a hard thing to dispose of without big equipment. While you may be able to butcher a lamb or goat on your farm, large pigs and cows generally require the use of a butcher. This requires the ability to both load and transport your animals to the butcher—a truck and livestock-hauling trailer. So while they yield more meat than chickens, rabbits, and other smaller animals, cows and pigs require more complicated infrastructure to get them from field to fork, and they may end up costing far more per pound even with their lower feed and other costs.

We have kept almost every common homestead animal—chickens, ducks, rabbits, pigs, and cows so far. (We have also helped tend and care for goats on a few different friends' places and learned that goats are not for us; if you think pigs test your fencing and your patience, you don't want goats in your life.) They all have benefits and drawbacks. If you are new to homesteading, it is best to start small, even if you go with a large animal. Don't jump into a bunch of cows. Get at most a pair to start. Don't start with a hundred chickens. Try out a dozen or so first. Raise a pair of pigs. Get your feet wet instead of risking drowning. Remember, what is easy in spring and summer—when most people get their first animals—is a challenge or, worse, a catastrophe in winter.

Pigs and compost make a good pairing—the pigs turn the compost while the compost provides food and warmth for the pigs.

Our kids' favorite and a wonder of nature, turning inedible biomass into delicious dinner options.

Harvesting Nature: Free Food of the Four-Legged Kind

In terms of cost, the best animals on the homestead are those nature provides, if you have the tools and skills to harvest them. Deer and elk. Wild rabbit and squirrel. Pheasants and other fowl. Wild hogs and pigs. In some areas, bears are now so numerous that conservation agencies are asking hunters to help control the populations. Hunting can help feed a family affordably and also benefit the environment by helping keep various animal populations in balance and check. Hunting is a dual good, helping both the environment while providing for and protecting those who eat them and those around them.

 ## Little Animals, Little Problems; Big Animals, Big Problems

My dad used to always say to me, "Little kids, little problems. Big kids, big problems." I can't vouch for how accurate this is as parenting advice (I swear my three-year-old is more destructive than a typhoon when unattended), but it certainly applies to animals.

Early one morning, as I approached the deadline for finishing this book, I received a desperate message on social media. It was from a dear family we have been friends with for many, many years. A few years back, they got into hobby farming. Bought a bit of land. Started keeping some backyard chickens. More recently, and the occasion for the call, they had acquired a half-dozen heritage breed cows set to calve a few months later. The half-dozen became a dozen as the calves hit the ground. Then, summer turned to winter. Winter got rough. They had to move the cows to a farm closer to their home. The cows got out. They got them back in. The cows got out again. And again. They ended up chasing the cows multiple miles in the middle of freezing weather with their entire family knocked down by the flu. They decided it was time to get out of hobby farming. They needed a hand relocating the cows as quickly as possible, which is how I got involved.

Realize, these are smart people with lots of resources who still struggled to make the jump from small backyard animals to modest hobby farming. Partly, it was just too big a jump to make starting off, and they both realize that now and say as much.

So learn from others. Start small. Don't overdo it, and don't overextend yourself. In the end, you may well end up with nothing but lost time and treasure on the endeavor.

My region is overrun by deer, who, because of their great numbers and lack of predators, cause a lot of damage to crops and cars. The damage to cars from collisions with deer sometimes also results in significant damage to the people in the cars. Other areas have problematic populations of feral hogs. Hunting them is an infinitely preferable form of population control compared to mass poisoning and other forms of removal that waste or contaminate the carcass and meat. That poison can easily impact other animals in the ecosystems, including yours and mine, and especially guard dogs. If you don't hunt yourself, remember to be supportive and appreciative of *responsible* hunters. They play a critical role in helping protect healthy ecosystems for us all.

Getting into hunting is no small thing. A good rifle and ammo will set you back a few hundred to almost a thousand dollars. If you didn't grow up learning firearm safety and skills, you are looking at investing a modest amount of time to develop them. Plan on three months of weekly practice under the supervision of a skilled hunter to get you up to speed, perhaps longer. Don't skimp on safety. Firearm accidents are all too common, even among experienced hunters and owners.

Bow hunting costs a bit less and is safer for the first-time hunter, but takes longer to achieve sufficient skill to hunt successfully. If you are going to live out in the country for the rest of your life and also have kids you want to pass such skills on to, then the investment will be more than worth it. Just keep in mind, it will take many years to recoup the cost, time, and equipment needed to learn.

So What Is the Best Animal After All?

Where does this leave us? In the end, the ideal homestead source of meat will depend on many factors. What is your land like? What is your infrastructure like? What skills and experiences do you have? What animals do you like to raise? Which ones naturally roam your land and are in need of a helping hand to keep their populations in check? Animals are an expensive investment, so start small. Choose one kind of animal at a time, gain experience, and as you learn what you like and don't like, continue to grow and expand. Remember, the bigger the animal, the bigger the risks, both economic and otherwise. Don't underestimate the need for and cost of fencing and heavy equipment necessitated by larger animals, or the time and skills it takes to safely and successfully hunt them.

Learn Plant Propagation

Plant propagation is a lost skill for most modern Americans. Perennial plants and plant starts can be very expensive, adding hundreds of dollars to a small garden's annual cost. Did you know that many common vegetable seeds are the same price per ounce as silver?

The ability to propagate plants can turn a hundred-dollar investment into a thousand dollars of savings and a thousand dollars or more of sales in just a few seasons. Propagation involves skills that cover both perennials and annuals. Annuals involve learning to select, save, and start seeds. Perennials come back on their own, but increasing their numbers or getting them established in new spots sometimes requires a helping hand. A single comfrey plant left to its own will slowly spread a foot or so each year. In an hour of work, you can turn that single plant into two dozen more, spreading it across your homestead to dozens of new places it would have never reached otherwise.

Below are plants we regularly propagate. Many of these are also easy to sell come spring, so we often propagate a few dozen extra of each variety.

Easy-to-Propagate Perennial Plant List
1. Raspberries
2. Blackberries
3. Mints
4. Comfrey
5. Oregano
6. Thyme (Thyme is *slow* growing; as a fellow homesteader joked, "It takes a long thyme!" Once it has a well-established root system, it does slightly better, but it takes a while.)

7. Chives
8. Yarrow
9. Aloe
10. Lavender and Rosemary

There are many, many more perennial plants that you can propagate, but in our experience the ones listed require very little experience or equipment to do so successfully.

How to Propagate Oregano and Chives

To propagate oregano, take a well-established plant or patch and find a fresh runner spreading across the ground. Gently pull up the plant stalk and some of the root below. Using scissors or plant shears, cut at least two inches below the stalk with the root system. Try to damage the roots as little as possible when removing them from the soil. Carefully take the cuttings and plant them into two-inch planting cups. Make sure that the soil stays moist and the plants stay out of direct, strong sun for the first few days to week.

You can propagate oregano almost any time, but if you want to get it established in a new spot, make sure you get it in the ground at least two or so months before frost so that it has time to become well enough established to survive the winter. Mulching a new patch as winter approaches to help ensure its survival is a good idea. I like using a couple of inches of shredded leaf mulch.

Chives are a type of bunching allium (a genus that includes members of the onion/garlic family). Once you have a plant, gently use a knife or blade to separate off part of the clump, cutting down and through the soil to capture both stalk and root. Take this and place it in a two-inch or larger pot. Although slower growing than oregano, the chives should settle in quite nicely and in a few weeks you should have a newly established plant.

Free Propagation Supplies

For seed starting and plant propagation, you need something to propagate and start into. Some people use egg cartons for seed starting, and they are a great option for some seeds and plants, especially for a smaller-scale garden and really shallowly rooted plants that you can easily get into the ground. Root binding is a real problem for plants, though, so I am not a big fan of any method that may harm my plants' health and thus reduce their vigor and yields.

Every spring, lots of propagation stuff—plastic trays, plant trays, and the like—are discarded. Asking around can net you *all* the propagation stuff you need for a few seasons. This is far preferable to purchasing new plastic, saving you money while reducing resource consumption at the same time. We generally get two more seasons out of recycled stuff and then make sure it goes to the recycling center. Keep an eye out for larger propagation pots. These are great if you want to sell some of your extra, larger perennial plants at a much better profit margin.

Come spring, the same store that I get pitch produce from will have a lot of planting supplies that get discarded from their annual plant sales, so during that time of the year I double up and enjoy the benefit of collecting both at the same time. Freecycle and similar groups are great places to let people know you want their unwanted planting supplies left over from purchasing annuals and other plants. Raiding the

Most of the materials we used to propagate plants our first few years were free or purchased at end-of-season clearance sales for pennies on the dollar.

recycling bins and dumpsters of big box stores that sell plants is another way to secure needed supplies.

The one concern with reusing propagation gear is spreading plant diseases or possible insecticide and other agricultural chemical contamination. Many stores admit to spraying things like neonicotinoid pesticides on their plants. So far, we have had no issues with reusing plant propagation materials, but we are getting almost all of it from local growers and gardeners who purchase their plant starts from trusted farmers The few times we have received stuff from larger big box stores, we also had no issue, but if I were getting such materials regularly or in large volume, I would do a bit more due diligence to make sure they were safe for my plants and pollinators.

Truly Free Food: Foraging from the Homestead

Another easy way to save around the farm and homestead is to take advantage of all the free foods that most areas offer in abundance, if you only know what to look for and how to properly harvest them. On our farm, we have at least 20 wild edibles—seasonal berries of many species, acorns and hickory nuts in abundance each fall, various greens

 **HOMESTEAD HOLLER** ## Preserving the Harvest with Dan Adams' Dry Canner

I love mason jars. They are infinitely better than so many methods of storage. Completely mouse and pest proof. But they don't seal well, especially the half-gallon sizes. While you can water bath or pressure can with them, that doesn't help for herbs, spices, grains, nuts, seeds, and dried fruits. These will still suffer spoilage and decay from exposure to air and moisture present in the jars.

Don and Dan Adams found a way to fix this problem. They took a standard Presto pressure canner and turned in into a vacuum sealing dry canner. Depending on the size of the canner and jars, you can vacuum seal a half-dozen or more jars at a time. Even better, you can seal stuff in those large, half-gallon mason jars that otherwise often go unused, but are such an efficient size for large families!

You can check out the full tutorial for how to build one at steader.com.

Wild edibles often hide in plain sight but can become a reliable food source for those who know where and what to look for.

through spring and summer, mushrooms in the woodlands, and greens of all kinds in the pastures and other places. The surrounding food-shed easily multiplies this abundance many times over.

We also encourage native edibles to flourish in our farm management. We already have started to or plan to re-establish many species and varieties on our property—elderberries and hazelnuts, persimmons and pawpaws are just a few examples. A small investment can reap decades of food income and security. Annuals are great, but perennials and native perennials are significantly better. Also, the better established edible perennials are on your property, the easier hunting becomes as well. The fruit, nuts, foliage, and other biomass is attractive to many other four-legged friends that can help fill up your freezer.

The best part of wild edibles is they tend to not need the same kind of care as domesticated plants. Spinach is finicky to germinate, and bugs enjoy biting into it as much as some people do. Lamb's quarters freely flourishes and self seeds year after year, often producing thousands of plants across the growing season with no cajoling from us. It requires nothing but a bit of space here and there and also provides food to our animals—chickens, cows, and pigs—who eat this wild spinach with exceptional vigor.

Brassicas like cabbage and kale face all sorts of pest pressure, requiring organic pesticides and floating row cover to protect them. Nothing bothers dandelion greens. Queen Anne's Lace is a perfect replacement for parsley. A few good books will go a long way in enabling you to successfully forage. My favorites are written by Samuel Thayer, who has multiple titles that cover a wide range of edibles found across the United States and into Canada. Many places now offer free or low-cost foraging classes as well that are well worth taking.

Keeping Foraged Foods Safe and Sustainable

If you have ever wondered about the safety of foraged foods, especially in more urban environments, a few years ago some researchers looked into the issue. In what turned out to be really good news, foraged foods tested as clean or cleaner than modern grocery store foods, especially fruits, seeds, nuts, and similar plant parts. Tubers and roots tested the highest for contamination, so you should show greater care and concern if foraging these in urban or other possibly polluted environments. Greens were in between, with some testing worse than others, so they are another area to show caution. If you know that a spot's soil may have been contaminated in the past, be more cautious in the present.

In additional good news, foraged foods tested higher for both macro- and micronutrient levels than their store-bought counterparts. Many of my friends have built maps of their surrounding foodsheds, showing where all sorts of foraged foods are available, collecting everything from serviceberries and apples, mushrooms and greens, to acorns and ramps, as the year moves from spring to summer to fall.

There are two concerns I want to raise with foraged foods. The first is a reminder to forage responsibly. Overharvesting of wild plants, such as ramps and ginseng, can quickly decimate wild populations and degrade ecosystems. A good rule of thumb is to take at most one of three for leaf and root plants—nuts and berries are a different ballgame—and don't harvest from places that show signs of recent foraging by others. The second concern is mistaken identity—misidentifying mushrooms and a few other plants is no small mistake. Every year, many people are sickened by consuming what they think are edible wild mushrooms. But mushrooms are not the only wild food that can ruin your day. Certain berries and even some greens are quite dangerous as well. When learning to forage, pay attention to plants that have closely resembling relatives that are not edible. When in doubt, seek counsel. There are a lot of resources on foraging online. Get creative with Google. Don't presume before you consume!

Eat Your Yard Out: The Benefits of Edible Landscaping

Even if you have a small property in the middle of a difficult HOA (Home Owners Association) or similar organization that stops you from growing food, nothing stops you from turning your yard and landscaping into a life-giving food forest. For almost every ornamental plant there are edible substitutes that work in both city and countryside. The interesting leaves and coloring of kale make it a favorite for filling in all sorts of spaces. Serviceberries, raspberries, and blackberries for bushes. Chives make great border plants, as do many other herbs. Indeed, herbs and medicinal plants such as lavender, thyme, rosemary, and so many others can easily take up large swaths of landscaping space. Artichokes provide visual interest and have become a favorite in some areas for edible landscaping.

Want an area with an ivy-like covering? Why not grow sweet potatoes, especially Japanese or another variety with interesting leaf shapes? Elderberry and hazelnut add height to the landscape while also adding beauty. Need to go big? Substitute dwarf or semi-dwarf fruit trees for ornamental and sometimes problematic species like Bradford pear. And, of course, you can add lots of flowers to attract pollinators and predators to your place to control pest species. Even better, look for species that are also edible, such as borage, calendula, nasturtiums, and so many others.

If you cannot have an annual garden in your front yard, there is often nothing to stop you from having a super-productive space filled with edible flowers, herbs, and plants that "hide in plain sight."

If you are going to spend money anyway making your place pretty, why not also make it productive at the same time? At the very least, your landscaping should be focused on pollinators and food production. Your landscaping should not only be beautiful, but edible and medicinal as well.

Medicine

Growing food is one thing that most everyone agrees almost anyone can do. But can you really grow your own medicine? Absolutely. As anyone can see on our homestead and in our vehicles, we still make use of modern medicine and technology. However, we also grow a wide range of natural medicines and make use of other native plants that our foodshed provides. Let's take a look at just a few.

Grow Your Wellness: Truly Natural Medicine

While not all food is medicine, many freely growing things around a farm or homestead are. Elderberry syrup has been shown to be an effective treatment for colds and flus. Plantain grows in disturbed soil and is excellent for treating stings, scratches, and other such injuries. Echinacea is a lovely flower, attracting pollinators and providing a powerful component used in many herbal remedies. In certain dry, thinly soiled spots on our land, mullein pops up reliably, and it has a long history of use for respiratory and other health issues. Jewelweed is an effective treatment for poison ivy—and generally grows close to it, conveniently enough. With our high tunnel in operation, we can now even grow tropical plants such as ginger and turmeric, both of which have well-known medicinal properties, beneficial for everything from nausea to high blood pressure.

> Let your food be your medicine and your medicine be your food.
> — HIPPOCRATES

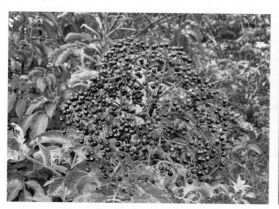 

Beautiful, medicinal, and beneficial to pollinators, elderberry is a must-have plant on any homestead if possible.

The list goes on and on. Many modern medicines are nothing more than concentrated extracts or knockoffs of what we freely find in nature. Purified and rendered more concentrated and potent, but still just isolates of what is available for free all around us or can be incorporated into our foraging and growing routines.

Some of these remedies also represent a significant savings if you make them yourself instead of buying them. Tinctures cost 20 or more dollars for just a few ounces. Many can be prepared at home for about a fifth of that cost. Elderberry syrup costs around $15–$20 for four to eight ounces. You can make your own for a quarter of the cost or less. If you have elderberries on your property and your own bees to provide the honey, it may run you less than a dollar.

Elderberry Syrup Recipe

This recipe makes approximately 16 ounces of elderberry syrup. If you end up with significantly more or less, this means you need to increase or decrease the length of simmering.

What you will need:

- 2 cups filtered or otherwise clean water
- ½ cup dried elderberries or 1 cup fully ripe fresh elderberries
- 5 whole cloves
- 2 cinnamon sticks or ½ teaspoon cinnamon powder
- 1 cup honey
- Optional—1 tablespoon freshly grated ginger or ½ tsp powdered

Put two cups (16 ounces) of water in a small pan. Add in all ingredients except honey. Cover with a lid and bring to a boil. Remove lid and reduce heat. Simmer until volume is reduced by half, stirring occasionally. Remove from heat. Strain out elderberries, cloves, and cinnamon sticks and toss these in the compost. Add honey to the mixture, stirring constantly to dissolve. Place into mason jars or bottles. (We prefer pint, 16-ounce, or 8-ounce sized mason jars.) Store in a cool place, such as your refrigerator.

Plantain for Stings

A simple way to treat stings and bug bites is with fresh plantain. Find the younger leaves on a plant, two or so, and chew them up. Apply the chewed-up plantain poultice to a fresh bug bite, especially bee or similar stings. (Make sure to remove the stinger first if it was embedded.)

Between what freely grows and what you can easily establish or add into your growing rotation, a homestead can offer you many truly natural medicines.

Cooking When it Counts

Now that you have all this food growing or you can forage or hunt to secure more, how will you cook it, especially if you are without power? Many years ago, the remnants of a massive hurricane moved through our part of Kentucky. Some places were without power for three or more weeks. Not hours. Not days. Weeks. Generators are great but can only keep so many things running, and fridges and freezers generally are the priority.

Wood stoves with cooktops, various types of sun ovens, and other such appliances are surely good to have as backups if you can afford them. Unfortunately, all of these have limitations, mostly weather dependent in nature. But something as simple as a few cinder blocks or a stack of bricks can be turned into a suitable cooking setup in a pinch. Even better, a lot of them use fuel that isn't useful elsewhere, such as smaller-diameter branches and debris, to generate heat, preserving your nicer and larger-diameter wood for warmth and winter. Let's take a look at some great options for the homestead when you need to cook and don't have all the conveniences of modern technology.

Four-Cinder-Block Rocket Stove

Rocket stoves use various marginal fuel sources—small-diameter sticks and other such pieces of wood, corn cobs, animal dung, and any other overlooked or discarded sources of carbon—to create lots of heat for short periods of time, which is perfect for cooking food. While we

This rocket-style stove lets you cook outdoors at little expense.

most often use wood for fuel, being able to use many other marginal fuels makes rocket stoves a versatile way to prepare food without expensive inputs or setups like propane, grills, or other gizmos and gadgets.

If you need even more cooking surface, with two additional blocks, you can create a six-block rocket stove with two burning spots. Also, for under $10 worth of bricks (regular or fire), or for free if you can find used ones, you can create a brick-style rocket stove as well.

Not All Bricks Be the Same

Not all bricks are the same. You probably don't find that surprising. But I am not talking about color or style. Instead, I am talking about durability. Certain bricks are specifically made for dealing with large amounts of heat. They are called fire bricks. So note that if you are planning on something more permanent for cooking, you should upgrade to fire bricks. Regular bricks and cinder blocks will deteriorate over time, though should still last you many, many months, if not more, for emergency or occasional cooking. If you want a permanent, self-sufficient way to cook, consider building an earthen stove or similar outdoor cooking setup.

Tire and Window Solar Cooker

In the hot Kentucky summer, it is sometimes nice to move cooking outdoors where, instead of heating up the house and costing us money running even more AC to deal with the additional heat, we can make use of the high outdoor temperatures to save on energy costs. Many homesteaders make or purchase solar ovens for this reason. Unfortunately, these are still kind of pricey and generally space restricted—we can't cook a full-sized meal for our family of seven in any common model.

One low-cost solution only requires a few spare tires and a window or two—the tire solar cooker.

Place a small sheet of plywood or flat sheet metal on the ground. Cover it with some aluminum foil. Stack a tire or two on top—if you need more space or have a larger cooking vessel or short tires, you may need to use two tires. Place your food in the cooking vessel and the cooking vessel inside the tires (black cast iron or similar dark cookware is best). Place the glass window on top, and let it bake. Just as a closed-up dark-colored car quickly becomes unbearably hot, this little set up can make reheating or cooking a snap. You will need mostly sunny to full-sun conditions for any solar oven to work, and the farther south you are, generally the more successful these sets will be. Even if a solar setup like this won't fully cook human food in some climates, it is a great way to reheat food, or to precook animal and worm food, allowing both to make better use of these foodstuffs.

You can find a number of helpful additional resources and options online.

The Instapot, Paper Plates, and Ensuring Your Sanity

Sometimes, it is worth spending some money to ensure or preserve your sanity. Paper plates, crock pots, Instapots, and other time-saving kitchen gadgets are one way we do this. Some days, things go wrong. You are in town longer than you anticipated. Or you spend two hours chasing pigs or hogs back onto your place after they find a breach in the fence and decide to spend the day seeing the surrounding countryside. Or you just need a down day, with fewer dishes and less stress, to enjoy some time as a family or get some needed rest.

This is another area I suggest you don't skimp on. If we want ourselves and our kids to enjoy homesteading the rest of their lives—city style, country style, or any style—then we don't want to burn out long before we finish building our bright dreams. That requires slack in the system. It requires opportunities to rest, relax, and take a step back.

So we have a box of 500 unbleached paper plates and two Instant Pots, along with a freezer full of easy meal options—sausages and hot dogs, pastured lunch meats, and easy sides. Make sure you make room for when things go wrong. Or for when you want things to go right and do something special, surprising, and spontaneous.

Tech Can Help Take the "Tough" out of Tough Days

Every year or so we add a piece of equipment to our kitchen that Jessica considers revolutionary. The first was a grain mill, which still grinds along to this day, turning low-cost organic foods into a myriad of delicious sides and dishes for our family. The second was a hand-held blender, which has been indispensable for soups, smoothies, and many other such creations. Each of these is great, but the Instapot—that has been a real game changer. Crock pots require preplanning, and when you have five kids and a farm, well, sometimes planning ahead just doesn't happen. Or all your plans go awry. In an hour an Instapot can make an entire meal, sometimes even from frozen. During long spring and summer days, the value of making a meal in just a few minutes of prep work and less than an hour to cook is beyond measure. The Instapot lets us make use of our preserved and fresh foods without having to spend hours in the kitchen or heat up the house. Given how time is often at a premium at our place, especially during the growing season, getting a few hours back while not bringing high-cost, low-quality food into the house is a win for everyone.

Also, the Instapot lets you make use of some of the lowest-cost or toughest-to-cook cuts of meat to make delicious meals without much fuss. Whole chickens, ribs, racks, shanks, and the like all fall at the feet of the Instapot.

Realize, homesteading is hard work—preserving, planting, building, splitting, and the thousand other chores you will always need to be doing. Don't hesitate to invest in making a few of these a little easier so that you and your crew can go about enjoying your homestead as much as possible.

Water

Many homesteads don't have access to city water. Some do have wells, but those often rely on electricity to work. If your electricity goes, it all goes with it, including your water and septic. Having other water options is crucial to providing for your family and livestock. One of the easiest ways to do so is to catch rain. It is generally abundant (albeit not predictably so). It is usually free (though, unsurprisingly, some cities and counties are trying to figure out how to tax it). It is also generally very clean compared to other water sources.

A thousand-square-foot roof in most of the United States catches thirty to sixty thousand gallons of water each year. An average house has about two thousand square feet of roof space. Most barns or outbuildings add another two or more thousand square feet of roof space to the homestead. Just a few buildings can provide hundreds of thousands of gallons of water a year if we can catch and collect it.

Here is a basic rainfall rule of thumb—every thousand square feet of roof produces a bit over six hundred gallons of water for each inch of rain. So if your area gets around 30 inches of rain per year, a 2,000-square-foot roof would produce about 36,000 gallons of water.

A simple flexible downspout lets us capture thousands of gallons of water in this water tank next to our barn each season.

City, Well, or Rain Water?

First, if at all possible, you really want more than one source of water for your homestead or house. Don't put all your eggs in one basket. If your place doesn't have a well, or city water already in place, in many areas it is now less expensive to do rain catchment than install a well or get on the water grid. While our well has never run dry, a number of our neighbors' have and have had to be rebored even deeper than before, adding a few extra thousand dollars.

When a friend of ours built their new place, a well would have run them around $8,000–$12,000. Rain catchment for their new place, totaling three thousand gallons of buried cistern storage and all the other components (including a high-volume filtration unit), was less than $4,000. With the difference in cost, they upgraded to a much nicer roof, copper gutters, and premium gutter guard to keep the water coming off the roof cleaner, reducing wear and tear on the entire system, and they still saved money.

Rain has a number of advantages over city and well water. You won't find chlorine or fluoride like in most city water supplies. No unpredictable price hikes. And no hardness like well water, which can ruin pipes, plumbing, and clothing, while staining toilets, sinks, and showers.

Rainwater catchment's biggest problem is drought. This is why the more cistern storage capacity you can afford, the better. The average city family uses about 60–100 gallons of water *per person per day*. Other estimates place an average family at twelve thousand gallons per month. Country folks tend to use less—but only if you don't include what you need for growing food and keeping animals—especially in summer, when the task of fertilizing and flushing can be rolled into

HOMESTEAD HOLLER Rain Catchment, Roof Composition, and Rendering the Water Safe

When it comes to catching rain, not all roofs are equal. Shingle roofs are made with a host of petro- and other chemicals, which end up in the water that the roof sheds. Slate, metal, and other roofing materials, while generally more expensive, produce safer starting water. Because of particulate matter, possible pathogens from bird and other animal droppings, and other debris, even a high-quality roof will require water filtration before the water is people- and house-piping safe.

Just like there is an array of roofing materials to choose from, there is an even bigger array of ways to filter water. At the very least, a sand, carbon, or other particulate filter should sit in between your catchment or cisterns to keep them clean and create an immediate reduction in debris and other problematic organisms having an opportunity to proliferate in your storage tanks. As the water enters your home, a UV, ceramic, or other filter can ensure that no dangerous organisms make it to your bodies.

one convenient, "go find a tree" reminder to young kids and adults. But even if you lighten your water load, a family will use six or so thousand gallons of water a month, or about two hundred gallons per day. At three thousand gallons of storage, that is about two weeks' worth of water. I personally would want no less than two weeks, with preferably a stored water supply of a month if my homestead's only source was stored rain water. So think of six thousand gallons as the minimum storage capacity you need on hand to ensure things go smoothly. If your farm is larger, or your dry season longer, you will need more.

If I was solely reliant on rain catchment, I would want about ten thousand gallons of storage for our family and another three thousand for our garden and animals. This three thousand also takes into account our pond, which contains another six to twelve thousand gallons as additional backup, especially for our animals.

The good news is that water storage generally scales down in cost— larger cisterns don't run that much more than smaller ones, so doubling your storage capacity may add only a small percent to your final tally.

Again, it is best to have multiple options for water, but if you can have only one to start, I would start with rain if your location and climate make it a reasonable first choice. If not, then you will need to see if well water is an option or if trying to get connected to the grid makes the most sense.

Once upon a time most houses and barns in the country had cisterns to store rainwater. It is a practice worth bringing back.

The Great Outdoors Calls

If your family has three males, fertilizing your farmscape instead of sending nutrients down the urinal would save 30 or more gallons of water a day (while also recycling nutrients to where they belong). If you have older toilets, the savings are significantly more—50 to 100 gallons a day. I mention males because, let's face it, it is generally far easier for us to go outside than our lady friends and requires no special equipment. So take one for the team guys, and make your farm more fertile at the same time.

On a few occasion, this backup setup has saved our bottoms!

Sawdust Toilets

A few years back, it was minus 30° for three straight nights in Kentucky. During the day, I don't believe we broke zero. Kentucky isn't made for such weather. Our water froze solid the first night. Not optimal, but we had bottled up a bunch of backup water, knowing this was likely. On the afternoon of the second day, our septic froze. No toilet of any kind for anyone in our then six-member family.

Fortunately, we had two things that helped us get by in such a situation—sawdust and a few five-gallon buckets. Put a few inches of sawdust in the bottom of the bucket. Make your deposit. Add a few more scoops of sawdust. A pile of sawdust isn't a bad thing to have around, and generally going into winter I fill a few spare trash cans with nice, dry sawdust in case of emergencies like this.

If our house made it feasible and fiscally sensible, we would move to waterless, composting toilets in a heartbeat. If you are starting from scratch, or retrofitting an existing place, especially one that has limited water resources, it is worth considering doing likewise, depending on the costs and benefits. Toilets account for up to 30–40 percent of the water a home uses per day. You can radically reduce your water needs and increase your water resilience by not wasting it on waste disposal.

Catching Rain with IBC Totes

There are a lot of uses for IBC totes. We've already covered many of them. But few match their value for rain catchment. Rain barrels are not overly efficient, unless you have a tiny garden and little other need for stored water. A good barrel is $20 to $50. At best, it will hold about 55 gallons. An average roof produces that much water in three to five minutes of heavy rain. Barrels can be daisy chained, but I don't recommend it. It is difficult to do, makes them unsuitable for other applications later, and creates risks of leaks, especially if you live somewhere with lots of freeze/thaw cycles. Along with increasing care and maintenance, it also substantially increases the time and cost to set up rain catchment.

An IBC tote costs $50–$100 and holds up to 350 gallons. Two side by side allow easy moving of the downspout when one reaches or nears capacity. Or they can be slightly offset and an overflow line run from the higher to the lower tote. Three IBCs hold over a thousand gallons of water. At that size, other water storage options are going to run two thousand or more dollars new, almost ten times the cost per gallon.

IBC totes do have a few downsides, especially if you are using them for your house water instead of just irrigation and animals. First, they tend to grow algae pretty aggressively. You can paint them or build something to block sunlight from reaching the tanks to help inhibit algal growth. Second, they freeze quickly in temperate to cold climates, so most people use these as seasonal water supplies, especially for outdoor applications like your growing spaces. In a pinch, you could use these year-round for supplying a home, but I would recommend spending a bit more and putting in true cisterns if rain catchment is your sole source of water. Otherwise, IBC totes are absolutely a great way to supply your irrigation and animal water needs especially during the seasons of heaviest demand.

Many houses have four downspouts, so placing two totes at each would result in almost three thousand gallons of water storage. I suggest lofting the totes if at all possible to create greater pressure. This is also why good home placement helps—a house should be on higher ground to minimize flooding and other problems. If placed properly, it also means water caught at the house can easily be gravity fed to other places where it is needed.

This can be done with some spare concrete blocks. Just make sure the setup is level, stable, and slanted slightly away from the house and towards the drain spout so that any overflow moves away, rather than towards, your home and its foundation or basement.

Swales: Keeping Water Where it Belongs

Few things add beauty, utility, and resilience to a landscape like a well placed and maintained pond. It can provide a backup source of water for irrigation and animals, a place to cool off or get clean during warmer weather, and habitat for all sorts of beneficial creatures in and around your land. Unfortunately, not all places can accommodate a pond.

 Getting Your Water Storage Right

A few months back, an acquaintance asked for some input on catching water off his house for use on his homestead. Underneath a deck, he had a nice, long, empty, and unused area where he planned to place six to eight IBC totes. He would then divert water from his roof gutters to fill the totes. Unfortunately, his idea had all sorts of practical problems.

First, the ground underneath the deck sloped *back* towards the house. So, if anything in the system sprang a leak, his house was going to pay the price. Hundreds to thousands of gallons of water gushing against a home's foundation is not a good thing. Next, he wanted to daisy chain them all together. He lives in a more northerly climate, and the area under the deck already tends to be windswept and cold. All the pipes and fittings and joints would be subject to a great deal of thermal pressure during cold stretches, possibly breaking, cracking, and then, when the big thaw came, leaking. There were numerous other problems, like the composition and stability of the soil. The area appeared somewhat unstable and prone to compaction, especially under the weight of multiple tons of stored water. The settling of the soil would be another way that the daisy chains could fail. Put it all together and it was problem after problem just waiting to happen.

If you are going to catch rain, make sure your placement and system don't create bigger problems than they solve. IBC totes when full are very heavy. They can easily sink into and compress soft ground. Three hundred gallons of water doesn't sound like a lot, until you spring a leak and it flows into an unwanted place, like your basement. Take the time to think through your setup and do it wisely.

Sometimes it is the slope of the land, sometimes the soil or subsoil or some other factor that prevents its inclusion. If at all possible, put ponds in appropriate places. If you don't know what good placement for your place would be, consult a qualified person.

If you are unable to put in a pond, you are not totally out of options for keeping water where it belongs. An easy and low-cost way to help keep water where you want it (in the ground and near your plants) involves building swales. Instead of having a flat run of land, swales involve gently undulating the landscape, providing places for water to pool and absorb into the ground rather than running off, down and away, causing erosion and other problems along the way. Swales allow water to slow, spread, and soak in—basic water-management goals. Properly built, swales allow you to grow more food with less watering. Combined with mulch and good soil, they allow us to grow crops with almost no supplemental irrigation in many years.

Swales do not need to be overly large or complicated. Ours are just a foot or two wide and half a foot or so deep along the upper side of our raised beds

Swales are a simple way to increase water retention and reduce erosion, protecting your soil while providing a proven way to reduce the need for irrigation. A properly constructed swale collects excess water during heavy rain, then slowly allows it to infiltrate the landscape, ensuring that your water can be caught instead of bought.

or some of our perennial plantings. Because our high tunnel sits above our main growing area, these swales help ensure that the extra, sometimes large amounts of water coming off the tunnel and into the garden are conserved and don't damage the plants and soil below.

Non-Electric Drip Irrigation

For watering plants, especially in a dry climate or inside a greenhouse or high tunnel, many growers make use of drip tape, electric pumps, and similar equipment. But what if you could use wood pallets, an IBC tote, and gravity instead to provide the pressure to the drip tape lines?

Four or five pallets stacked on top of each other, followed by the IBC tote on top creates sufficient pressure to run a number of drip irrigation lines. Another option if you need more pressure (each foot of elevation creates about 0.5 PSI of pressure) is to loft one IBC on top of another. Two totes reach a maximum height of about eight feet. Most drip tape needs a minimum of 3 PSI to work.

If you have access to sufficient used, free, or low-cost PVC pipe, you can buy joints and drill holes in the pipe to create an alternative to drip tape.

Not only does this save time and money while improving plant health, it saves water. Air watering, like with a hose or sprinkler, causes water loss via evaporation. Drip irrigation lets you get the same results for a fraction of the water.

Irrigro makes lower PSI drip tape, which is perfect for a setup such as this: irrigro.com.

Heat

People love chopping wood. In this activity one immediately sees results.
— ALBERT EINSTEIN

Cold weather is a far bigger problem than warm. People congregate around the world's equator but tend to stay away from the poles for this reason. Perhaps you are fortunate to live in a temperate region with a home that is easy to keep warm and cool. But even here in Kentucky, the so called South here in the States, we faced weather into the minus 30s a few years back, so cold even our septic froze. While that was a bit unusual, stretches around zero are fairly common each winter. So much for the South!

In such extreme cold, HVAC systems struggle to keep up. Many of our neighbors had to leave their homes or face heating bills into the mid-hundreds of dollars or more during that one bad stretch. Even moderately cold spells create community-wide grumbling over rapidly increasing utility bills.

For us, it takes just a bit of extra wood to get through a hard few days or a colder few weeks. Wood is really the best answer in a lot of places for an affordable, alternative heat source. Personally, I dislike pellet and other stoves that use alternative fuel sources that I can't produce or easily procure myself. These resources also reduce access to wood chips, sawdust, and similar inputs in some areas. Since these are so valuable to farmers and homesteaders and they are already hard to come by because of their demand as fuel sources, I will stick with making use of real wood that, if called for, I can cut, split, and stack myself right on my farm.

Easy Firewood Options: Cut-Offs, Slab Wood, and More

There is an old saying: "Wood heats you twice." I found out over the past eight years of homesteading that that old saying is wrong. Wood heats you six, sometimes seven, and even eight times. When you cut down the tree. When you delimb the tree. When you cut the trunk of the tree and large branches into rounds. When you split the rounds down into wood. When you load the wood to move it to the stacks. When you unload and stack the wood. When your stack falls over and you have to restack the wood. When you move the wood into your stove.

There is no more efficient method of heating the human body and losing weight than heating with wood on the homestead. A single cord of wood produces enough heat over the course of the year to make saunas superfluous for us homesteaders. It also represents a tremendous amount of work. A cord of wood weighs between four and six thousand pounds. Generally, you will move it three to five times before it is finally burned. Thus, each cord of wood may well run you ten to twenty thousand pounds of handling labor. We go through a measly four or so cords each winter, yet we feel that fifty to eighty thousand pounds keenly. Some of our friends farther north go through ten or more cords, a truly Herculean feat weight wise.

Old wood
best to burn,
old wine to drink,
old friends to trust,
and old authors
to read.
— FRANCIS BACON

Slabwood that isn't too barky makes great low-cost firewood for use during the milder parts of late fall and winter.

Cutoffs are also great for cool weather and protecting your cord wood, while also providing a great deal of easy kindling.

Firewood takes a lot of skill and a fair number of tools to do safely and efficiently. Even though our family has achieved both the skill and tools—appropriate chainsaws, safety gear, and a hydraulic wood splitter—over the years, we still like some of the low-cost firewood options that our area presents—slab wood and sawmill cut-offs.

Slab wood is the outer wood, bark and all, that comes off a log as it is turned into rough-cut lumber. A bundle (about three thousand pounds or so) sells for $5 from the local sawmill in our area. The mill has an entire acre of bundles just sitting there. Even at that rock-bottom price, they can't get rid of them. If you have an outdoor wood-burning stove, slab wood is often one of the best low-cost options, if available in your area.

The other wood source we really like is rough-cut dimensional and pallet cut-offs. The same sawmill cuts wood for pallets and custom dimensional wood orders. Every day of cutting the sawmill produces a big pile of leftover pieces that are either too short, have bad spots, or otherwise don't make the cut. For $5 they dump the entire pile into the bed of my pickup truck. Often, not only do we get a good bit of firewood from a load but lots of usable rough-cut pieces of lumber, which you have seen us make use of throughout the book. The pins we make for our raised beds come from these loads. The cross pieces are used for propagation tables or scrap pieces for cobbling things together as well. Also, the same scrap pieces can be used to make wood racks that store wood better.

A Simple Wood Stacker

Our first year homesteading, a good friend showed me this simple design for making wood stack end holders for both drying and storing firewood. They take just a few minutes to create, and they hold up well and use the weight of the stack itself to hold secure the ends of the wood pile. You can vary the height of the stackers based on your needs.

Material List

- ▸ 2×4s or similar wood
- ▸ Screws or nails (I prefer screws, but the originals were built with nails)

Dimensions

- ▸ 4 pieces cut to 30 inches
- ▸ 2 pieces cut to 10 inches

For the piece lengths, this is a basic setup, but you can vary the width of the stacker and its height and length based on the size and type of firewood you use.

These sit nicely on top of pallets and act as excellent "bookends" to support the piles. If you are stacking wood on pallets, you can fit two rows, but leave ten or more inches between the rows to ensure sufficient air space to allow your wood to properly dry.

Another simple rack design a fellow homesteader came up with uses the following:

- ▸ two 4×4s or two 2×4s for the horizontals
- ▸ four 2×4s for the uprights
- ▸ 3 cinder blocks

This design requires no screws. While some people omit the center cinder block, I don't recommend it, as the horizontal boards will bend and warp badly over time, making them less reusable and more likely to fail. An extra cinder block is a minor cost for the added security of not having your pile end up on the ground, so I recommend you keep it. If you are spanning more than ten feet, I would place a cinder block every four feet or so for the same reason. A 12-foot stack would thus require four blocks—two at the ends, and two four feet in from each end.

Another reason to provide this extra support is for the safety of those working in and around the stacks. A stack of falling firewood is a dangerous thing. So I want to be sure my kids, animals, and guests are safe when bringing more wood inside or just passing by the piles. An unstable pile of wood is an easy way to get someone hurt or just create a mess when it topples over and requires restacking.

The High Cost of Free

One important lesson to learn is about the cost of free. A while back, a friend messaged me on Facebook about having some wood he wanted to get rid of. I asked a few questions, trying to ascertain if the wood was in good condition. It appeared it would be. So I loaded up my kids, hitched my trailer, and headed 40 minutes or so over to his place. I immediately realized that much of the wood in the pile was rotted. After an hour or so of loading, we headed home, where we had to clean off rot from a lot of the wood for another hour or more. Only then could we stack the much reduced pile. (If you are looking for a good use for rotted wood, see the potting mix recipe in Chapter One.) Four or more hours for probably less than two ricks of good wood, plus the cost of driving to and from his place. It was a net loss.

 ## Putting Newspaper to Good Use

I don't get the newspaper and don't think much about it. But you may find a few stacks of mags and rags around our house. They serve an important purpose—newspaper knots for starting fires. Instead of spending money on fire starters or using chemicals or other accelerants to get fires going, years ago we saw this technique and took time to master it. Seven years of wood burning later and we have yet to be disappointed.

Take a few sheets of newspaper and roll them into a tube shape. Next, tie the tube in a knot. Make about six knots or so. If you don't have good kindling, make more. Make your wood stack. Toss a bit of kindling on top of the pile. Toss in the knots. Toss a bit more kindling on top. Light it and enjoy a great fire in just a few minutes.

These make an excellent replacement for other types of firestarters and are free.

I don't fault my friend in any way. It taught me an important lesson—free or cheap can be costly. Ask lots of questions, request pictures, and do all the due diligence you can before you take half a day (or more!) on something that isn't worth the effort and expense. Your time and a tank of gas are not free. An extra few minutes asking questions or requesting they inspect the items first pretty much are.

Storing Wood: Leftover Greenhouse Plastic and Pallets

Wood requires careful seasoning and storage for many reasons. First and foremost, your life and the lives of those in your home depends on properly seasoned, dry wood. Each year in my area many houses and lives are lost or ruined by chimney fires. Such fires are usually caused by creosote buildup, which is the result of burning insufficiently or improperly seasoned wood.

Improperly seasoned wood is also inefficient wood. Remember all that time and energy you used building up your wood supply? Look at the big pile of wood that you are so proud of. Wet, green wood burns with about a third to half the BTU (heat) value of properly seasoned, dry wood.

So take half that pile and toss it in the rubbish pile. That is what

Woodsheds can be made from a wide range of materials. These sheds, on the Mitzel and James homesteads, each serve them well and allow ample wood storage.

you are doing when you burn wet wood. Proper seasoning and storage aren't that hard and can be done at relatively low cost. Like with hay and straw, stopping ground contact is crucial. Stacking wood on pallets or rails is low cost and also speeds up seasoning, since air is able to flow through and under the pile, while also protecting it from ground moisture. An added bonus is not having to bend so far to get the bottom pieces of your pile and risking injuring your back in the process. A pallet shelter for wood is even better.

Also, wet wood is waiting wood...At least you are waiting, waiting to finally get warm once all that excess water is burned off. It takes a *long* time to finally get a house warm with wet wood. Need heat in a hurry? You will be fighting with wet wood for hours before your house finally heats up. Take the time to properly season and store your wood. Let's show you how we do so affordably.

If you have a true woodshed, that is great. It is definitely a worthwhile build on most homesteads. But if it is out of budget or low on the priority list, these options can help you get by.

Tarp Trick

As I've already mentioned, I don't like standard store-bought tarps, neither the regular nor heavy-duty varieties. In our area, I find they don't last long enough, and they puncture far too easily to be of any use protecting a wood pile. Also, with all the wind we have, they tend not to stay in place easily, and the grommets give way if you try and use those to secure it. I generally tell homesteaders to save their money for something more durable or take advantage of other ways to protect their wood.

But if you already have a heavy-duty tarp and want to try a nifty trick, here it is. Choose a tarp large enough to double over your pile. It should be a foot or so off the ground on each side (I prefer my tarps to not sit on the ground, since this can lead to freezing and then tearing of the tarp material). Along the edge with the grommets, use some rope to tie each set of grommets together. You should now have a double-layer tarp with two seams. In the seam, place a length of extra metal or plastic pipe for weight. With plastic pipe, if more weight is needed, you can fill it with gravel and add end caps. Two people can now easily lift up the tarp to access the wood, but wind and weather won't blow it off the pile.

Community

A lot of homesteaders move out of the city to pursue a more self-sufficient lifestyle. But without community, you will never achieve self-sufficiency. Honestly, there is really no such thing as "self"-sufficiency. Even Ma and Pa Ingalls had to head to town on occasion (and what a trip that could be) and relied on the help of neighbors, even if "neighbor" meant someone three miles away. Instead of pursuing self-sufficiency,

 **Matthew Eby's Quick Metal-Covered Wood Stacks**

Another option to keep wood dry for those without woodsheds and with too many other priorities to take the time to build one is using old roofing metal sheets to cover rows of stacked wood. If you are using wood cut to around 20 inches in length, the sheet metal will sag down on both sides a fair bit, offering extra protection. For well-seasoned wood, we stack it on used pallets two rows deep by as long as we have space for. Since we are cutting wood about 18 inches long for our stove, a standard 36-inch width of sheet metal will adequately cover a tightly stacked double row. We generally make the stacks four feet tall and hold down the metal with large rocks or cinderblocks to keep the metal from flying away.

Credit: Matthew Eby

Sheet metal can make an excellent cover for seasoned firewood.

 Adam Barr's Rocket Mass Heater in his High Tunnel

In many places, greenhouses need supplemental heat to protect or successfully propagate plants. It is often pretty pricey to keep these structures sufficiently warm, both in terms of the equipment and then the energy to power it. A farmer friend of mine, Adam Barr, wanted to go a different route. His place has lots of firewood and he wanted to figure out a way to use this free source to heat his high tunnels. Our area also has lots of clay, so a solution came together quite easily—a clay-based rocket mass heater.

The problem with most approaches to heat is that they heat air, which has low thermal mass and retention. The hot air doesn't last long, requiring more energy to heat more air in an endless cycle. A rocket mass heater heats a large mass (generally earth, stone, or water), that then radiates heat for many, many hours to come. This is far more efficient than many other methods of heating and far less volatile. The mass heats up more slowly, stays warmer longer, and gives off more even heat for a number of hours after the fire has gone out.

Adam hopes to upgrade his heater this coming year by adding more thermal mass in the form of water to further extend the amount of time the system can go between cycles.

This rocket mass heater allows Adam to germinate and protect plants during late winter with excess wood from the farm, rather than having to use electricity or other, more expensive, heating options.

> Do not forsake your friend and your father's friend, and do not go to your brother's house in the day of your calamity. Better is a neighbor who is near than a brother who is far away.
>
> — PROVERBS 27.10

our goal is to achieve *community* sufficiency. To be rich in relationships that benefit everyone around us.

Honestly, as a homesteader, there really is no other option. Who is going to take care of your animals if you want to go on vacation? Well, you might say, homesteading is my vacation. I won't argue with that (though I think you probably will want a vacation one day, anyway). But what about if a family member far away dies and you need to attend the funeral? Or you or your wife get injured or sick? Or have a new baby? Life is not easy, even in the city. When you add the demands and burdens of a homestead, there is no way you can do it without help. "Better a friend close by than a brother far away," says the book of Proverbs, and for good reason.

I can't count the number of times neighbors have helped us catch or close up animals, checked on things when we needed to be away for a few days, helped us with projects, let us borrow tools or equipment, and much more. In return we have let them hunt on our land, lent our stuff, picked up their packages, and tried to be neighborly in whatever ways we could in return. Sometimes that meant grabbing some extra gas to say thanks for letting us use the riding lawn mower. Other times

HOMESTEAD HOLLER — Jessica on a Mission

When you have five kids and a farm, clothes can become a fairly substantial expense. Kids are hard on clothes. Homestead and farm kids are by far the worst. In many rural areas, there are stores like our local one (which is called the Mission). Similar to a thrift store, they are full of all sorts of deals on all kinds of stuff. For us, the best deal is brown bag weeks. For four bucks, you can fill an entire grocery bag with clothing. You can fit a *lot* of kids' clothes in a grocery bag. For under $20, Jessica often gets about half of

the clothes our five kids need for about half of the year.

The Mission, and a few stores like it, along with our church family, has been crucial to us affording clothes for our big family. We make sure to give back and hope others will do the same. We check first with our church to see if anything we no longer need but that is still in good condition would serve another family, and if not, it goes back to the Mission to find its way to another family in our area.

it involved giving a few pounds of our freshly butchered pigs to say thanks for all the help over the past few months. Water, heat, and food are great. But don't forget community. Without it, none of the others matter.

The Promises and Perils of Sharing Equipment

One way to save large amounts of money is to share equipment. For the first five or so years of our homesteading adventure, we co-owned both a log splitter and a dual-axle trailer with a friend. Both of us at the time wanted and needed both pieces of equipment. We had a clear understanding that if you damage it, you fix it. General maintenance and repairs were a shared expense. This saved both of us a few thousand dollars and allowed us to have better equipment sooner than we could have otherwise as well.

Many self-sufficiency people think that they need to own everything to succeed, but the opposite is actually true. Expensive, occasionally used equipment, like trailers, log splitters, and a hundred other things, tie up limited, valuable resources and capital and actually slow you down from achieving self- and community-sufficiency and resilience.

If you can't find someone to share with, in many rural areas neighbors have equipment that they will let you borrow, barter for, or rent. I used to need to haul a few hogs or cows to the butcher on occasion. A trailer would set me back at least three to four thousand dollars, plus insurance, upkeep, and more. Instead, I asked a neighbor up the road if I could rent his. He offered it for 25 bucks for the day. Given I needed it at best two or three times a year, it was a major deal for me, and some money towards new tires for him.

Borrowing from others requires you to have a good reputation for properly caring for other people's things. Do you return the truck with a full tank of gas and a tidy interior? Do you get something borrowed back on time, or does it sit under the side of your barn for months on end? If you break something, do you let the person you borrowed it from know and then take care of all arrangements for a full repair, not just throw a few dollar bills at them and leave them the burden of getting it sorted out? Be a giver, not a taker. Go the extra mile.

While renting was a big savings, a year or so ago I found an even better option. Another farmer friend goes past not too far from my place on the way to take his animals to the butcher. So I just pay him to toss mine on as well and save both the time, gas, money, and aggravation of having to haul a few animals a year hither and thither. Given that I used to lose half a day or more taking a few animals to the butcher, on top of the rental cost, gas, mileage, and everything else involved, paying him 50 to 60 dollars to move a few pigs is a deal, probably saving me two to three times that amount.

With another farmer, I bartered worm castings for high tunnel construction help. He got much needed fertility, I got a 2,100-square-foot high tunnel put together with a lot less physical labor as well as access to equipment that would have cost me a few thousand dollars to rent and many, many thousands to purchase.

Maybe the best way to put it all is this: Relationships and community are far more valuable and cost effective than all the equipment in the world. There will always be another tool, another implement, another piece of infrastructure that you need on occasion. Owning them all isn't feasible. There will always be another skill that you need—electrical, plumbing, small motors, framing, and the like. Among a few families, you can get a lot more covered at a fraction of the cost.

My road has a mechanic, a metal worker, a nurse, a plumber, a metalsmith, a farmer, many skilled hunters, and who knows what else across just ten homes. Together, there isn't a whole lot we couldn't do or help each other with if circumstances called for it.

This really comes down to an irreplaceable homesteading and farming lesson. Stuff is great, but skills are invaluable. If you can hunt, fix, build, heal, forage, and the like, you can survive in a lot of circumstances. But if you can get by only if you *buy*, then your homesteading dreams are going to the by and by. If you can take your skill set and barter it for other skills and stuff you need without having money enter the equation, you'll do more than survive, you'll likely thrive.

Combining Pieces and Projects

As a child of the 1980s, I grew up with (and on) cartoons. I always loved the ones where the heroes could individually combine their vehicles or powers to make something better, something greater, something more powerful. The Dinobots in *Transformers* or the lions of *Voltron*. So it is with what you have learned in this book. The various pieces and parts are not ends, but beginnings—things that you can combine and recombine to create new, better, more fitting infrastructure, animal housing, feeding and watering setups and so much else for your farm or homestead. Let's just highlight a few.

Pallets Plus Cattle Panels: Andy Buchler's Animal Shelter

Want an animal shelter that has a bit more height? Randy Buchler used pallet walls plus cattle-panel arches to create a much taller animal shelter that is great for larger creatures you may keep. It also makes it easier for people to attend to birthing, injured, or sick animals, or otherwise work in and around them, since no slouching, squatting, or other special postures are required—you can stand and work normally.

Credit: Randy Buchler

Randy Buchler combined pallets, cattle panels, and scrap sheet metal to create size-appropriate housing for different animals.

Instead of cattle panels, you could use the trampoline conversion to create such a structure, running the metal tubing down into the space provided by the upright pallets.

Sheet Metal or Greenhouse Plastic + IBC Tote Cages for Wood Storage

If you are using IBC tote cages to store firewood, how will you keep rain off? You can use sheet metal or greenhouse plastic. This works even better than trying to lay either of them directly on the wood piles. With metal, the unevenness often causes issues and the need for restacking every few weeks, while with plastic, contact with wood eventually causes holes, tears, and punctures.

You can cut plastic to cover a single tote of wood, or cut a longer piece to cover mul-tiple totes stacked side by side. If the wood is still seasoning, I would stack the totes in a line, not a square, as the wood where totes meets will be awfully slow to season and dry out properly. If stacking in a line, I would leave a minimum of six to twelve inches between totes during seasoning.

Simple Outdoor Storage

We often have piles of materials or wood that we plan to use in the coming weeks, so instead of sending them all the way up to the barn, far from where they may be used, we make temporary storage out of wood pallets and sheet metal. This allows us to stage materials for projects when we first get them in, instead of having to move them multiple times.

Credit: Dave Perozzi, Wrong Direction Farm

The Perozzi family combines IBC tote cages and pallets with sheet metal to store easy-to-move stacks of wood.

A large batch of almost free number 2 cedar lumber from my neighbor's sawmill, which was used for our high tunnel, was staged just a few feet away and protected from weather by a few pallets and sheets of scrap sheet metal.

Speeding Up Wood Seasoning

One final idea to show how to combine projects. If you want to season wood at a greater speed, or if you struggle to get wood properly seasoned in a more humid location or climate, you can use the trampoline greenhouse as a passive solar wood-seasoning shed. Inside, you can place multiple IBC totes of wood; the typical time to season may be reduced by half or more. Or you can use pallet or other racks described in this book to fill the shed and keep the wood up off the ground.

The only change I would suggest for this application is adding some venting as high up as possible in the shed, preferably on both end walls. A few companies make vents specifically for use with greenhouse plastic. Also, the shed should be located in a sunny location that's as dry as possible, especially not some place where water tends to pool or run.

You could also make the end-wall plastic easy to pull back during warm or sunny weather but easy to move back into place during rain or snow.

Caleb's Composite Bow

With some leftover pex pipe and poly rope, Caleb created his first bow, using sharpened sticks for arrows. With help from dad, his siblings soon made their own upgraded versions with real practice arrows, and were off playing and hunting.

[6]

Sourcing and Resources

IT's EIGHT O'CLOCK on a Friday night. Friday is buying club day and we often pick up stuff for a friend or two who don't live close to town but want access to good-quality food from the club. So we often get some visitors later in the evening as people head home from work and swing by to grab their stuff. Robert, our friend and neighbor from a few miles down the road, rolls up around nine o'clock to pick up his family's order and visit for a few minutes.

After a few pleasantries, he asks me to put on my boots and come outside to his truck for a moment. What does he show me in the bed of his truck? A beautiful selection of boxes, crates, and other items from his work's discard pile. Hundreds of dollars of lumber and hours of assembly, all sitting right there, ready for reuse.

Robert mentions that his work tosses a number of these *every week*. All sorts of notions dance through my head. Nesting boxes, storage bins, raised growing beds for herbs, pantry storage, tool storage. (Note, by the way, they are HT stamped, so they are heat treated, and also have no stains, so they are good for just about anything you can think of .)

Stuff like this abounds. Empty three- and five-gallon buckets from food businesses. Steel and poly barrels from larger food preparation kitchens and operations. IBC totes from large industrial or food operations. They are out there. You just need to learn how to find them. Let's give you some pointers.

The modern economy produces lots of valuable resources that often go straight to the dump unless someone takes the time to repurpose them.

Every time I drive around, my eyes see so much opportunity for resource reuse and cost savings.

Construction Sites and Construction Supply Stores

During large construction projects or construction booms, a lot of good materials go to waste. The big caveat with visiting constructions sites is ensuring you don't break the law or otherwise get into trouble. Some places may not allow you on site. Even if they do, if you accidentally take the wrong stuff, it may be the most expensive scrap 2×4 you ever did snag. So while this is a great resource, make sure you are mindful and clear on what you are allowed and not allowed to take.

Every construction supply store or site in my area has large roll-off dumpster that all sorts of stuff goes into. They don't mind me taking a peek to see if anything worthwhile is salvageable, since the faster it fills, the sooner they have to pay for its removal. This is a substantial expense, so every pickup load of wood and other materials salvaged is another 30 or so dollars for their bottom line. They also have stacks of standard-size wood pallets they love to see people take off their hands. Out in the country, there is no pallet recycling like in larger cities that sometimes cuts into the supply.

For large wooden pallets, hitting up specialty stores is the path to success, especially roofing, sign making, and similar operations. Unlike smaller pallets that pallet recyclers sometimes gobble up around cities and make hard to find, these larger pallets are too big for most people to make use of or handle. The large size is what makes these even more useful for farmers and homesteaders, allowing you to make larger fences or structures far more quickly than with standard pallets.

Libraries

This is going to strike you as strange advice from someone you just bought a book from and who owns probably a thousand-some books or more. We have so many books that I think bookshelves are our most plentiful piece of furniture and books our most numerous personal belonging. But here goes: Don't buy books! At least, follow the advice I was given in seminary and as a general rule: "Borrow before you buy." This doesn't apply to my books or books I recommend. You can trust those titles to be worth having in your home library!

For years of college and higher education, I bought dozens and dozens of books that I would never read or use again, usually available only at full retail cost. Buy at retail, read once, and then sell for pennies on the dollar. Many, in just a few short years, were worthless, replaced by new editions or no longer used by instructors. It wasn't until my third year of master's work that a professor friend gave me this great advice. "Borrow before you buy. Ask friends, use the library, read the free e-version. Once you have read a book, you will know if it is actually worth owning it."

 ## Boom, Bust, and the Big West

One of my friends is from a ranching family out west. Thus, they are also in energy country—oil and natural gas abounds underground. When talking about farming a few years back, he mentioned they rarely have to buy fencing or a lot of other supplies for their very large farm. Whenever new oil and gas wells are put in, once construction is completed, the extra materials are generally left piled near the site. It isn't worth it to the rig crews to try and haul the leftovers all the way back to their base of operations or to the next site. Rolls of fencing, pipe, posts, and so much else just turns to rust and then dust unless someone comes along who wants it. My friend says you have to ask, but often that is all it takes to get a big pile of brand new, high-quality building materials.

Our farm and homestead book collection contains some two to three hundred titles. In a given year, we probably consult about 30 to 40. Another few dozen I would want on hand no matter what—reference volumes or other great works for if and when they are needed. But that leaves a hundred or so books that just sit there, collecting dust and taking up space, and that, honestly, we may never look at again our entire lives. A few are still worth having, as we lend them out regularly to others, sometimes to never see them again.

With libraries, if you dislike the selection you can do something about it. When I became frustrated with the lack of quality titles at our local branch I began taking time to suggest books and movies to the procurement people. Libraries generally purchase books that they think people want to read. So if you and others in your area get involved and make suggestions, you can change their perspective on what they should be shelving. (Make sure you ask them to carry my books while you are dropping other suggestions.)

 **HOMESTEAD HOLLER** ## Whole Life's Lending Library

For side work, I do book reviews for a few different publications. This means in a given year I receive a few dozen new books, beyond any that I personally purchase or otherwise procure. While a lot of them are great, most are not something I will ever need or want to read again. Not because they are not good books, but the subject matter or other material just isn't applicable to my current life. The constant flow of books was overwhelming our already overburdened book shelves and our house bursting to the seams with five kids.

On the other hand, the buying club we started years ago in Louisville has lots of people who would love to have access to such great titles, especially since local libraries rarely carry the kinds of books I generally review—books on food, farming, and alternative health. So the club started its own lending library. Dozens of families donated other great books along with ours. One member set up a free online checkout system for the members to use to check out books. Now, hundreds of books have new life, and our community has yet another benefit to attract new members.

Perhaps your church or another social group can do likewise and everyone can benefit.

Craigslist

I have found some amazing deals on craigslist. I have also wasted a lot of time scrolling around on it as well. A few of my best finds included a 1,500-gallon water tank setup for just $300 and a $4,500 wood stove with everything extra included for $600. To succeed on craigslist, you need to be able to get to good deals quickly before others do—the good deals don't last long. Really good deals are generally faster than a lightning strike.

If you are close to or in between a number of major cities, or looking for items where a few hours' drive would still be worth it if the right deal comes along, SearchTempest allows you to search multiple craigslists at the same time. Also, at some times of the year there is a lot less competition than others. Wood stoves fall in price come spring, when people aren't thinking about the coming year's winter, and jump up in price in fall. Remodeling and renovation supplies go up though during springtime but tend to go down in the fall and winter, when fewer people want to do that kind of work on the inside of their homes.

You Have Not Because You Ask Not

When I started looking for a wood stove, I didn't realize all the other stuff that you need to install it. A hearth pad. Heat shields. Flu and chimney. Roof supports and so much else. Once you factor all these in, a wood stove can run many thousands of dollars, even used. When I found a deal on a stove on craigslist, I went and put a deposit with the seller while I gathered a crew to help me get the five-hundred-pound stove safely to my place. After we removed the stove itself from his home, I noticed that the hearth pad, flu, chimney and other stuff was just kind of there, in the old owner's

This ended up changing from a floor to a through-the-roof deal with a simple question at the time of sale.

way. So I asked if he wanted me to take all of that as well. His response? Well, sure! What would have cost us over fifteen hundred dollars was now part of an already amazing deal. Thirty minutes later it all joined the stove on the truck and trailer and off it went to its new home.

A lot of times if you buy something, it may come with a lot of other accessories, pieces, tools, or parts. Even if the ad or offer doesn't mention or include them, ask after you have agreed on a price. At worst, they'll say no. At best, you'll get them for free or, more likely, at very low cost.

Freecycle

In some areas, Freecycle groups may be worth a look. I personally haven't had much luck with them so far, but some of my other homesteader friends have, especially those in urban or near urban areas. Partly, it is because we live far out from the city and these groups tend to favor those closer in. Freecycle groups used to make use of Yahoo groups, but some are also on Facebook now as well or have moved exclusively over to Facebook. These groups require that you give before you take. Don't expect to hop on and start getting stuff until you have taken time to give a few things first.

Facebook and Facebook Marketplace

In my area, a number of groups have popped up for people to barter or give away unwanted stuff. They are like Freecycle, just moved over to Facebook. There are also a lot of sales groups, some solely for farm and homestead type stuff. Each group has an easy-to-use search feature—in really active groups, don't bother trying to scroll through endless postings hoping to find what you need. Also, Facebook has rolled out a keyword feature that automatically notifies you when a post with that keyword (T-post, lumber, etc.) is made. This makes it very easy to find wanted items without wasting a lot of time. You can set a bunch of keyword tags and wait for Facebook to tell you that something popped up.

Some of my homesteading friends have said that Facebook Marketplace and sales groups have almost completely replaced craigslist and similar person-to-person sales channels in their areas. Others have said the opposite, that Facebook Marketplace isn't very active in their area. Try to figure out which works best in your area and region.

Friends

One of the best resources I have found is friends in particular lines of work. If you have to purchase barrels or totes through resellers, you will often have to pay a major markup of two, three, even four or more times what buying direct from the source will run you. If you can get these items directly from the business, the savings is immense. They may sometimes even be free. That becomes a lot easier if you have connections. Let your friends in certain lines of work know what you are searching for. If they help you hit the jackpot, remember to say thanks with something from your farm or homestead that they will appreciate.

Barter Makes Friendships Better

I have acquaintances who have access on occasion to great resources, Not people I know well enough to get something for free from or impose on heavily but whom I could easily barter with if I have something they are looking for or want. For instance, years ago, the husband of a lady who frequented the buying club worked at a place that produced a large amount of discarded lumber each week. These were solid oak 2×4s around 12 to 16 feet long. He just happened to be a hunter who was always on the lookout for the next buck. Our homestead just happens to be a deer super-freeway system in need of constant traffic control. So he brought me a large load of lumber every time he came out to hunt on my land. Free lumber plus farm pest removal is a deal I couldn't pass up!

One of my neighbors works for a roofing company. When they put new roofs on, old ones come off. When scrap metal prices are low, there is little incentive for them to recycle the old metal. It actually costs them more to take it to a metal recycling center than give it away. So for a few bucks or a chance to hunt on our place, he will often run me an entire load of sheet metal worth many hundreds to thousands of dollars new. Sometimes, the stuff is in such good shape you could use it on a house if you lined up the holes properly. For chicken and rabbit tractors and outbuildings and all sorts of other projects, it is perfectly acceptable, especially at a 95 percent or more cost savings.

I am still reusing lumber from five or so years ago and sheet metal from three or more years ago today. All secured for a few bucks or a simple barter of something that didn't cost me a dime—letting them hunt on our land—and actually helps out our homestead.

Salvage and ReStores

In some areas, salvage and recycle stores now help keep stuff out of landfills and back in action. Some are better than others. Those that re-condition and then resell barrels, IBC totes, and similar business waste at times may significantly overcharge for these items. In my area, the IBC tote resellers are two to six times more expensive than direct sales. Animal mineral totes that are free from neighboring farms cost ten dollars a piece. IBC totes that are as low as $25 go for up to $150.

Some, like the Habitat ReStores and similar building seconds and supply outfits, offer a lot of great deals if you know what to look for and where good prices fall.

Pawnshops and Similar Businesses

In some areas, pawn shops are numerous, but their prices are generally unimpressive. Sometimes you can find good deals on tools. Sometimes the prices, given the condition of the tools and lack of any kind of warranty, represent a gamble at best. If you are going to shop pawn shops and similar establishments, make sure you are familiar with the retail prices, especially the "on sale" retail prices of what you're looking for. Sometimes you will find a brand-new item on sale for not much above the used price at a pawn shop. All the same, I have gotten a few very good deals over the years, but not so many that I frequent pawn shops very often. They are an occasional kind of event, if I have a few minutes to spare, or if I am looking for a particular thing that I know they will have at a reasonable price.

Salvage and Tear Down

Something has to eventually happen to old barns and buildings. Sometimes, insurers want them removed for safety reasons and will pay to have someone do it, especially after damaging weather events leave outbuildings and other structures in a state of danger. Other times the owner will give you access to take something off their place that is unwanted and will eventually become a problem if not dealt with. Sometimes it is free, sometimes they ask for a low fee. A few of my friends now specialize in such demolition work and thus end up with a lot of free material for their farms and homesteads. If you can't do this work yourself, you may be able to find someone locally who does it and strike a deal to get the materials through them. If they are not using them, they have to dispose of them, which is a cost both of time and waste disposal fees. So many may be inclined to strike a mutually beneficial deal.

Estate Sales, Auctions, and Farm Auctions

The average age of a farmer in America is over 65. Rural areas in general have similar demographics. This means there are a lot of estate and farm sales and auctions as farmers move on to the hereafter and almost no one, family included, wants to take up their pitchforks or shovels.

Such sales offer a lot of opportunity for beginning homesteaders to get their hands on all sorts of tools and supplies for pennies on the dollar. Trailers, hand tools, water storage, old-style food preservation items like mason jars, lumber, farm implements, and so much else pop up at these sales and auctions. In some areas, the auction companies maintain websites with dates for upcoming auctions and a glimpse into what is for sale ahead of time. Some auctions allow you to drop by a few days before the event and see what will be available. This is helpful, since you can go in with an idea of what you want and what it's worth before the heat of the auction happens.

Don't overlook low-cost items at these events. Partly used boxes and cans of nails, screws, and so much else often go for next to nothing. Stacks of lumber often go at 80–90 percent savings. These odds and ends are often overlooked by other bidders but can save you a bundle on stocking up your homestead reserves. If you plan to bid on bigger things, make sure you have the ability to pick them up within the terms and limits of the auction. Sometimes, that means same day.

There are many websites to help you hunt down auctions, such as auctionzip.com. You can also use the telephone book or internet to search your area for auction companies, which often have their own websites where they list upcoming auctions. Craigslist and a few other places also can help you track down upcoming events.

Online Auctions

You no longer necessarily need to attend auctions in person. Over the years, I and some of my friends have scored big with online auctions. For us, it was an $8,000 walk-in cooler for around $800. Other times it was getting wanted toys or other items for the kids (or the big kid) from Goodwill's online auction. Many homesteaders and farmers have gotten excellent deals on all sorts of kitchen and butchering equipment and infrastructure from restaurant and business auction sites. While

 ## To Tractor or Not to Tractor

Everyone wants a tractor. It seems that most people finally feel they have arrived when a tractor is taking up space in their barn or some other building. But I wanted to mention something about tractors and similar equipment. For new homesteaders and part-time farmers, these things are immensely dangerous. Around the US, over the past ten years or so, as homesteading and farming have regained a place of appreciation, a surge of accidents has accompanied them.

Heavy equipment is mighty expensive, not just to purchase but also maintain. Years ago I realized that, given the cost of a good tractor, I could barter for one or rent it from a neighbor for the few times a year when I needed it and save a significant chunk of money in the process. If you do get bigger equipment, realize that older models often lack the safety improvements of modern stuff. They are generally easier to maintain but more dangerous to you and your kin. Make sure you get sufficient training, and always stay aware and sober minded when using it. Something as relatively small as a large rock in a field can cause a tractor to roll over, endangering anyone on board. Be safe, friends!

online auctions allow almost anyone anywhere to bid, be sure you understand the rules and factor in the travel or shipping costs to collect your winnings. A great deal on a two hundred dollar item disappears if you lose an entire day to mileage and travel. Overnight stays to collect stuff require very large savings to make up for the costs involved.

Even the government has an auction site that a few of my friends have reported success sourcing stuff from. See what's available near you at govdeals.com/.

Thrift Stores

One clear winner in our area is the local thrift store. While we haven't found a lot of stuff directly helpful to projects, the ability to buy a grocery bag of clothes for five kids for four dollars is a pretty big savings to a homesteader. Bike helmets and similar gear are often just a dollar or two, compared to ten to twenty dollars brand new. Some outdoor deals include nice planting supplies.

Roadsides and Parking Lots

Strange as it sounds, we have found tons of stuff just sitting out along the road or next to dumpsters at apartment complexes. One day when we were driving home from church, a house that was undergoing renovations had a two-hundred-dollar brand-new stainless steel shelving rack right next to the dumpster in the driveway. I went and knocked on the door and asked if the person needed it gotten rid, and he responded, "Yes, please take it." We have found bicycles and toys for the kids set out next to people's trash. Firewood just stacked along local roads by people who took down trees but don't use wood to heat and want it gone but don't want to see it landfilled.

In some cities and counties, they have set days for specific zip codes or neighborhoods where people can put out larger items for disposal. Again, consider your time and fuel investment, but sometimes it is as simple as taking a slightly different route to and from your normal stops and shops to save hundreds of dollars and find some incredible stuff.

Your Local Landfill or Recycling Center

Depending on your location, some local dumps will let you pillage and plunder the detritus and debris that is brought in daily. A few friends mentioned that sometimes to get access you need to give the attendant a few dollars. If you can scope things out first, it can be a few dollars well spent. This is especially true of satellite locations rather than the main dump.

Automated Online Shopping

While we try and make do, reuse, and buy local, there are still many things we get online. As I mentioned above, automation helps a great deal with trying to get certain items, like on craigslist. A few other helpful sites for online shopping are Honey and CamelCamelCamel.

Honey lets you know if you are getting the best possible price at that time. CamelCamelCamel lets you see historical price information on an item for places like Amazon and set price alerts for when something drops to a point you want to purchase it at.

My kids have really wanted a swimming pool for a few years now. The model that would work best for our place and budget retails for around four to five hundred dollars. Last winter, I saw it for around $270, but then it suddenly went up to over $400 before I could buy it. I was bummed, as were the kids. I didn't want to miss such an affordable price again. So I set some alerts and then went on my way until just last week it dropped to a bit over two hundred dollars. So for just a few minutes' work we will get a great swimming pool at almost half its normal price. Even better, this is about what most people in our area are charging for a used one.

One of our favorite homestead knives goes on sale once or twice a year at around half off on Amazon—it makes a great all-occasion gift as well as a "keep one everywhere and anywhere" knife, like in vehicles, medical kits, up in the barn, and more. At their normal price, they are still a great deal, but when they go for half off, it is too good to pass up. But who wants to spend lots of time checking for such sales? Automation means that every time the deal pops up, we are in the know.

A few other good services that are popular among my friends, both homesteaders and regular, are Slick Deals and Mass Drop. For things like computers and other tech gear (this book wasn't written on a typewriter), Slick Deals is hard to beat. They keep tabs on the best of the best deals and sales from across thousands of websites and sellers. Mass Drop is an excellent way to get everyday carry (EDC) and other gear at a great savings, including things like heavy-duty socks, flashlights, head-lamps, and so much else. "Every day carry" refers to items that every person would be wise to keep on hand but especially homesteaders—a pocket knife and multi-tool, a flashlight and headlamp, a water bottle and medical kit.

A final online option worth mentioning is the new quarterly or monthly box services, such as Battlbox. These companies charge a flat fee and send subscribers a box of selected tools and gear at a deep dis-count over retail—generally 50 percent off or more. While the savings some months are substantial, subscribers have no choice over the con-tents, so such a service isn't for everyone, and some months may not provide things you want or need.

Services like these also come and go, so I try to find and become friends with someone who can keep me in the know about which ones are worth keeping an eye on and sales that may be of interest to me.

Acknowledgments
(A Final Homestead Holler!)

If I have seen further
it is by standing on ye shoulders of Giants.
— Isaac Newton

⚬——⚬

If I had to footnote and name all the farmers and people whose ideas made this book possible, it would take two more volumes. As Isaac Newton said above, at the end of the day all of us stand on the shoulders of others. Perhaps two ideas in this book are truly my own. To homesteaders everywhere, from generations back to those still plugging along today, my thanks for inspiring and enabling our family to flourish.

My thanks to all the homesteaders and farmers who helped contribute great ideas and projects to this book. So you can look them up and thank them yourselves, if so inclined, or find out more details about ideas and projects in this book, their websites, blogs or other resources are given below. Many have great sites, blogs, and other resources that I encourage you to check out.

To those who read the various stages of the manuscript draft, my thanks as well. Your input in terms of quantity, quality, and organization will hopefully benefit every poor soul who buys this book.

My many thanks to the following friends for contributing to this book:

▸ The Aud Family

▸ Adam Barr
  Barr Family Farms

▸ Randy Buchler
  Shady Grove Farm   facebook.com/ShadyGroveFarmUP

▸ Carlos Cunha
  The Five Cent Farm   Carloscunha.org

▸ David Good
  thesurvivalgardener.com

▸ Matthew Eby
  Eby Farms   ebyfarmsllc.com

▸ Jordan and Laura Green
  J&L Green Farm   jlgreenfarm.com

▸ Brian Gallimore
  Gallimore Family Homestead   gallistead.com

▸ Chris Hollen
  Hollen Family Farm   hollenfamilyfarm.com

▸ Dave and Rachel Perozzi
  Wrong Direction Farm   wrongdirectionfarm.com

▸ Jeremy and Rachel James
  Acres Wild Farm   acreswildfarm.com

▸ Travis and Janelle Veldkamp
  Homestead in the Holler   homesteadintheholler.com

Final Note

I HOPE YOU HAVE ENJOYED the book and all the various contributions people made to make it possible.

If you want to stay in touch and see my up-to-date recommendations on resources, books, supplies, and more, check out my website, homesteaderhandbook.com. Whether it's links to savings on products or suggested tools and suppliers, my site will have more time-sensitive information to offer.

You can also catch some of my classes and talks at Steader.com or other places online. Or perhaps we will cross paths at a Mother Earth News Fair or some other conference or event in the coming years.

Whatever way we connect, please know how much I appreciate your support, hope that you enjoy the book and any other material I create, and that I can't wait to one day connect in person.

John, Jessica, Abby, Caleb, Noah, Ellie and Lydia Moody.

Thank Yous

Before we finish, I wanted to say thanks to a few people who I am deeply indebted to. First, Jessica, my amazing wife who didn't have any idea what she was getting into when she said, "I do!" Who knew five kids, a farm, and so much else was in store in these short 14 years? Here is to many more!

To Ingrid, Murray, and rest of the New Society team, my deep appreciation for all your work. To a few friends who helped proof and correct early chapters and improve and refine the content, and make sure it was accurate, thank you. I know a publisher when you all are finally ready.

To Joel, Teresa, Daniel, Sherri, and the rest of the Polyface Farm crew, thank you all for your friendship, and Joel for your constant encouragement and counsel. Sorry it took me so long to get moving on some things, but better late than never.

Index

About the Author

John Moody is a homesteader and farmer and is founder of Whole Life Services and CEO of Steader, an online video-based learning platform that brings the best practitioners to guide and inspire those who find the joy in raising and growing their own food. John speaks and presents at homesteading conferences and events and is a contributor to numerous magazines, journals, and websites. John, his wife, and their five children live, farm, and homestead on 35 acres in Kentucky.

ABOUT NEW SOCIETY PUBLISHERS

New Society Publishers is an activist, solutions-oriented publisher focused on publishing books for a world of change. Our books offer tips, tools, and insights from leading experts in sustainable building, homesteading, climate change, environment, conscientious commerce, renewable energy, and more—positive solutions for troubled times.

We're proud to hold to the highest environmental and social standards of any publisher in North America. This is why some of our books might cost a little more. We think it's worth it!

- We print all our books in North America, never overseas.

- All our books are printed on **100% post-consumer recycled paper**, processed chlorine-free, with low-VOC vegetable-based inks (since 2002).

- Our corporate structure is an innovative employee shareholder agreement, so we're one-third employee-owned (since 2015).

- We're carbon-neutral (since 2006).

- We're certified as a B Corporation (since 2016).

At New Society Publishers, we care deeply about *what* we publish—but also about *how* we do business.

Download our catalogue at https://newsociety.com/Our-Catalog, or for a printed copy please email info@newsocietypub.com or call 1-800-567-6772 ext 111.

New Society Publishers
ENVIRONMENTAL BENEFITS STATEMENT

For every 5,000 books printed, New Society saves the following resources:[1]

29	Trees
2,613	Pounds of Solid Waste
2,875	Gallons of Water
3,750	Kilowatt Hours of Electricity
4,750	Pounds of Greenhouse Gases
20	Pounds of HAPs, VOCs, and AOX Combined
7	Cubic Yards of Landfill Space

[1] Environmental benefits are calculated based on research done by the Environmental Defense Fund and other members of the Paper Task Force who study the environmental impacts of the paper industry.

new society
PUBLISHERS
www.newsociety.com

More Resources for Homesteaders

The Modern Homesteader's Guide to Keeping Geese
Kirsten Lie-Nielsen

7.5 × 9" / 144 pages
8 page color section
US/Can $19.99
PB ISBN 978-0-86571-861-6
EBOOK ISBN 978-1-55092-654-5

Raising Rabbits for Meat
Eric and Callene Rapp

7.5 × 9" / 208 pages
US/Can $24.99
PB ISBN 978-0-86571-889-0
EBOOK ISBN 978-1-55092-682-8

Raising Goats Naturally, 2nd Revised & Expanded Edition The Complete Guide to Milk, Meat, and More
Deborah Niemann

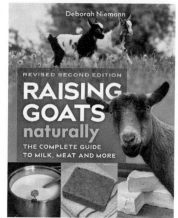

7.5 × 9" / 352 pages
US/Can $29.99
PB ISBN 978-0-86571-847-0
EBOOK ISBN 978-1-55092-642-2